CMOS IC LAYOUT

CMOS IC LAYOUT

Concepts, Methodologies, and Tools

Dan Clein

Technical Contributor: Gregg Shimokura

Newnes

Boston   Oxford   Auckland   Johannesburg   Melbourne   New Delhi

Butterworth–Heinemann supports the efforts of American Forests and the Global ReLeaf program in its campaign for the betterment of trees, forests, and our environment.

Library of Congress Cataloging-in-Publication Data

Clein, Dan, 1958–
    CMOS IC layout : concepts, methodologies, and tools / Dan Clein;
    technical contributor, Gregg Shimokura.
        p.   cm.
    ISBN 0-7506-7194-7 (pbk. : alk. paper)
        1. Metal oxide semiconductors, Complementary—Computer-aided
    design.   2. Integrated circuits—Computer-aided design. I. Title.
    TK7871. 99.M44C485   1999
    621.39'732—dc21   99-44934
                                                                   CIP

British Library Cataloguing-in-Publication Data
A catalogue record for this book is available from the British Library.

The publisher offers special discounts on bulk orders of this book.
For information, please contact:
Manager of Special Sales
Butterworth–Heinemann
225 Wildwood Avenue
Woburn, MA 01801-2041
Tel: 781-904-2500
Fax: 781-904-2620

For information on all Newnes publications available, contact our World Wide Web home page at: http://www.newnespress.com

10 9 8 7 6 5 4 3 2

Printed in the United States of America

*To my wife Emilia, who has put up with my hobby
of layout design for the past 15 years.
To my kids Noran and Nathan.*

CONTENTS

PREFACE

Once upon a time, around about 1988, after finishing a very stressful but successful project within Motorola Semiconductor Israel (MSIL), the entire team was invited to a special lunch. Everybody was happy that we finished the "project" ahead of time, and we were there to enjoy the victory of "tape-out." Instead of sitting in separate groups, IC circuit designers, CAD support people, and IC layout designers sat intermixed around round tables. I had the opportunity to sit beside Zvi Soha, who was at the time the CEO of MSIL. After enjoying a very special meal, but before the dessert arrived, Zvi asked each of us to tell him what would make each one of us more efficient, happier, and thus more productive. I list the various answers below:

The *IC design engineer* asked for faster workstations, more copies of the simulation software, and more engineers.

The *IC layout designer* asked for faster machines, place-and-route tools, more people, and better support from the CAD group.

The *CAD representative* said that all they needed were more and more people, because they wanted to provide Motorola with a complete software solution that would enable the CEO to "push a button and have a complete chip instantly ready." The idea was that if Zvi needed a new chip, the software would ask him to fill in the fields of a pop-up form with the required specification numbers, and pushing the "enter" button would result in the final design. The CAD representative went on to explain, "With such powerful software you will not need all these design engineers and layout people that were always asking for more software and hardware."

After a few minutes Zvi's answer was:

"Well, you know, if I have such powerful software, I will not need you (CAD) either. . . ."

The moral of this real-life story is that in the past decade, most people thought that with the help of very advanced and sophisticated software, all the major problems would be solved.

It is true that as the gate length of devices became smaller, the density of the chips increased, the design complexity increased, and the time-to-market requirements shrank, teams of designers had to find new ways of dealing with the many challenges.

What is very difficult for design automation partisans to understand is that by the time a new design automation tool is widely accepted, the challenges have changed.

For example, when block sizes and design complexities grew to a point beyond human capabilities to lay out manually, *floorplanners* and *place-and-route* tools were introduced to automate the layout process.

In the beginning these tools were driven by schematic-based design styles. But when the circuit complexity and size grew, CAD adapted and *synthesis* appeared.

The next step was to adapt the place-and-route tools to synthesis, and so on. . . . If we analyze the development of all automation software, we may find that all the development was driven by people who were ready to change, but who knew *why* things are the way they are and *what* they could do to change to find new solutions for the new problems.

Yes, automation helps—but the change and evolution in design was always driven by people who understood the basic concepts, tried new methodologies, and drove CAD software designers forward to develop new tools.

So it is under this umbrella that I will try to help all interested designers, both circuit and layout, and CAD developers to understand more about the real world of layout. That's why my book will talk mostly about concepts, methodologies, and tools related to CMOS layout design.

A few years ago at the Design Automation Conference, I was invited to participate in a demo of a new floorplanner. I was so impressed by the performance of the tool during a 10-minute demonstration on the trade show floor that I asked to see a private 40- to 50-minute demonstration.

In the same room there were about five people from different companies. The software developer was very proud of his remarkable tool and started to explain all about the features of the tool. For almost 30 minutes he amazed all of us with many screens full of options for floorplanning at different levels of integration. Everybody was impressed with the vast capabilities of the tool.

During the last 5 minutes we, the potential users, were invited to ask questions. The room was very quiet . . . everybody left fast, after only one very banal question was asked.

When I was alone with the developer, I had my own simple list of questions. I asked him the following:

During the development of the tool, did somebody think about potential users—who they were, and what their level of software knowledge was? Based on the number of things they had to set up, this was not an easy job. Assuming that people with limited software background will use the tool, there were 200+ fields that needed to be completed, and many others that were automatically set. Only then did you push the button and get an idea of the results. If more tweaking was required, then the driver of the tool would need to ask an expert for help or would have to learn the advanced features and capabilities of the tool.

The answer was, "We didn't think about this. . . ."

The sales pitch for such a tool should demonstrate more than just advanced capabilities. Ease of use was a critical issue that was overlooked!

I suggested that the development team should have had an advisory committee that is made up of a variety of potential users from different companies with varied requirements and methodologies. Did this happen in their case?

After a few more questions like this, I realized that in this case 20 software engineering Ph.D.s with very limited experience or knowledge about physical layout created a wonder of a tool based on a dry specification but without feedback or cooperation with any potential users.

This was another moment when I thought about this book. It is very difficult to design and build a tool for layout without knowledge about layout concepts and methodologies.

I am sorry to say that this "wonderful" tool is still *not* on the market so we the users can benefit from its capabilities (sorry, but no company names).

Similar things have happened to me many times over the years, so in this case I decided to give the tool developers a hand. Yes, we need better tools, but we have to help tool developers to understand more about our philosophy as users. At the same time, we as users have to understand more about the philosophy of the tool. When a tool is to be designed, the technical marketing department that generated the specification had something in mind, and the final tool should reflect this view.

Using new tools means that we as users have to adapt our thinking and our methodologies to accommodate the new tools. The best example to demonstrate this is the application-specific integrated circuit (ASIC) flow. Only companies that started from scratch or built groups based on the new flow and methodologies were able to survive the problems of changing the way to design with the complex and different tools brought on by the new trend.

A smaller initial capital investment than before is required and less expertise is needed to use these new tools, as an ASIC flow has enabled a great many new companies to enter the IC and system design marketplace.

Most big companies have internal training courses for all levels of design, internal CAD groups to develop design tools, and a lot of resources for research, but there are advantages to being small. You can adapt faster to the new trends, methodologies, and flows.

Without having the overhead of internal tool development programs, small companies have to be more creative in finding solutions with much more limited resources. Small companies have to adapt to the offerings of external vendors such as Cadence, Mentor, Synopsys, and Avant!.

Their tools are not built specifically for any of us. Instead, they reflect market trends more than any internally developed CAD tool. These vendors do not operate completely independently: if one company buys 1,000 copies of a software package and another buys 20, the first company's voice is considerably stronger for the vendor in influencing new features for the tool. There is always the threat of competition just around the corner, so there is still much more incentive to be right the first time. . . .

Let's briefly list the major challenges of an IC designer in CMOS today. I would have liked to call this preface the "umbrella" chapter, because the problems from one project to the next are like a heavy downpour, and I hope that my 10 chapters will help all of you to survive the flood.

PART ONE: THE BASICS

Where does layout design fit in the overall chip development process? Chapter 1 gives a nontechnical overview of the entire process so that we can understand the layout designer's role.

The mandate of an IC layout designer is to create the layout masks of various portions of a chip in compliance with engineering drawings, netlist or simulation results, and process design rules. To be capable of understanding and respecting engineering drawings, the designer needs to understand basic electricity rules and all the concepts related to the layout of gates. This will be covered in Chapter 2.

Chapter 3 describes the manufacturing process and definition of layers. After we understand how the layers are coordinated to generate devices and connectivity, we learn about design rules. These are the manufacturing rules that must be followed to ensure that the chip can be reliably manufactured. The process engineers determine the minimum manufacturing grid, polygon, minimum distance between layers, etc. The design rules are the rules that are the factor, which together with the engineering drawings, netlist, etc., will fundamentally decide the architecture of the chip.

PART TWO: LAYOUT STYLES

If a Layout Designer does not respect design requirements, the chip won't work. If the design rules are not respected, then the chip may not make it out of the prototyping phase. The art of a good layout designer is to combine both, while taking into consideration all the other aspects of a normal project: time to finish, final size, quality, and so on. . . .

None of the chips just mentioned can claim that they are made up of only one type of design style these days, so in Chapter 5 we talk about specialization in design. We discuss full custom, standard cells, gate arrays, and other types of techniques used in today's ICs and the advantages and disadvantage of each type. We talk about various techniques and methodologies used in complicated chips for specific applications. The list is long, but some of them are clock generators, datapath or register files, I/O cells, and memory types. We end the chapter with chip finishing techniques.

PART THREE: ADVANCED TOPICS

The topic of Chapter 6 is related to the requirements of big chips for adequate connectivity and power routing. We learn about methodologies to address all these and discuss placement impact to routing, floorplanning techniques and results, preplanned signals, etc.

Chapter 7 assumes that we know the basics and we start dealing with analog problems, such as capacitors, electromigration, and 45-degree layout, to mention only a few.

Special process requirements are explained in Chapter 8. Learning about slits in wide metals, step coverage, latch-up, and special design rules is possible now that we understand even the most complicated process rules.

When the environment is uncertain, meaning that the process is not defined yet or the design not 100 percent simulated, the layout designer has to face new challenges. That's why, in Chapter 9, we learn about contacts as cells, test pads, spare logic gates and spare lines, and laying out a circuit with changes in mind.

PART FOUR: TOOLS OF THE TRADE

Perhaps the most exciting chapter is Chapter 10. This chapter analyzes various EDA layout design tools required to face the challenges of any kind of layout design. From crude polygon generation to place-and-route, from generators and silicon compilers to verification tools, from plotting devices and software to transfer formats, we try to show you a path through this maze of names, concepts, methodologies, and usage. This chapter does not try to rate or recommend specific tools, but it does try to enlighten the novice user about the choices in the marketplace and how these tools might be adapted to different methodologies, and vice versa.

This book is intended to help you protect yourself in a downpour of complicated design methodologies pitched by EDA vendors, a world in which the names of companies and tools change all the time, the hot topic each year is different, and every year pundits at the Design Automation Conference are announcing new catastrophes and solutions.

For example, first the machine was too small (CALMA). Then UNIX came along and more memory was needed. Place-and-route appeared, along with verification tools, extraction tools, and new terms like Deep Sub-Micron (DSM), and so on. Even if the tools are solving most of today's problems the market requirements (prices) are always generating new "unsolved mysteries."

This book is meant to help you prepare to understand the basic and advanced concepts, and to learn how to analyze new methodologies and to understand the philosophy of new tools. I hope that it will be useful for all of you, and I will be more than happy to receive your comments. Please write me at the following address:

Dan Clein
826 Riddell Avenue North
Ottawa, Ontario
Canada
K2A 2V9
cometic@ieee.org

ACKNOWLEDGMENTS

Unlike any other book, this one is the product of people's communication and willingness to spend time and explain why things are the way they are. I have tried to list all the "contributors" who, over the past 15 years, helped me to learn and understand concepts, methodologies, and the tools used for layout. This book is not only mine; it is theirs as well, because these are the people who believe that teaching others will make their life easier and the companies they work for more successful. The list is in chronological order, not necessarily related to the importance or quantity of information that I received from them. Together with you, I thank the following:

Miriam Gaziel-Zvuloni—she was the person who saw potential in me and hired me as IC layout designer even though I barely knew Hebrew. She was the first teacher for all the basic layout I have learned. (INTEL—Israel)

Zehira Sitbon-Dadon—my manager for more than 5 years, who pushed me to learn and develop many advanced layout concepts. She offered me the opportunity to became the layout teacher, to manage projects, and be responsible for all the layout tools and interfaces with vendors, engineering, and CAD within Motorola—Israel.

Nathan Baron—the first circuit designer who invested time in teaching layout designers what, how, why, etc., engineers expect when designing a schematic. His favorite saying to any new problem was, "First let's sit, and slowly, slowly (relaxed) we will find a solution to any problem!" (Motorola—Israel)

Israel Kashat—the Director of Engineering who always helped by answering all the process questions by saying: "What a nice problem. It is good that we found a problem. If we do not find any problems and have to solve them, why will somebody pay us a salary?!?" (Motorola—Israel)

Steve Upham—a very enthusiastic Application Engineer who spent 5 months trying to promote new tools and methodologies within Motorola Israel, who explained to me in great detail the philosophies of symbolic editors and place-and-route tools for the first time. (Cadence—England)

Carina Ben-Zvi, Nachshon Gal, and Eshel Haritan—CAD people who worked with me to develop various internal tools for layout and many times had to explain software limitations, concepts, and philosophies. They often helped me to become better prepared to understand software developers from various vendors. (Former Motorola Israel employees)

Jean-Francois Côté—the first Canadian engineer who introduced me to DRAM layout secrets. His approach was then, "The more I teach others how to do what I know, the more time I have to learn new things . . ." I really believe that he is right. (Former MOSAID—Canada)

Graham Allan and Cormac O'Connell—my teaching experts in designing memories. They taught me most of what I know today about layout concept related to analog layout, DRC weird rules, and DRAM process requirements. (MOSAID—Canada)

Ed Fisher—being Mentor Graphics' "guru" in the IC Graph polygon editor, he enhanced my knowledge of the capabilities of such tools, including my first encounter with device generators. (Mentor Graphics)

Jim Huntington—the Cadence "guru" in verification tools who helped us learn, install, and successfully use DRACULA on 16-Mbit chips.

Glenn Thorsthensen—another Mentor application engineer who spent a lot of time with the MOSAID layout group explaining place and route and compactor tricks. (Mentor Graphics)

Michael McSherry—he is the technical marketing person who introduced me to hierarchical verification concepts and implementation. (Mentor Graphics)

Steve Shutts—the first software developer who explained more than the ROSE tool, he taught me how symbolic layout tools and layout synthesis can make a difference in an IC layout designer's work. (Rockwell)

Dennis Armstrong—a layout designer who moved to tool benchmarks and enhancements. For all of the past 10 years, he has helped me understand a lot about various tools. We began to talk while I was working for Motorola, and we continued to exchange tool information over the years. (Motorola-Austin)

Dan Asuncion—layout teacher for the Institute for Business and Technology (IBT), Santa Clara, California, who generously shared with me a lot of layout teaching experience and his course curriculum. He is one of the people who continuously encouraged me to write this book by promising me that he would use it as the reference for his classes.

Mark Swinnen—former Silvar-Lisco application engineer who helped me understand more about placers, routers, and analog and digital considerations in the place-and-route environment.

Ron Morgan—one of the owners of GERED Corporation who sent me without too many questions the curriculum of their training courses so I could base my Canadian IC Layout course on an established North American style.

Roger Colbeck—the VP of Engineering in the Semiconductor Division of MOSAID who gave me the opportunity to manage and build the first trained IC Layout Group in Canada.

Tad Kwasnivski and Martin Snelgrove—professors at Carleton University-Ottawa who encouraged me to come and teach VLSI students what the industry wants them to know. Being in front of students without any written training material pushed me to start working harder to write this book.

Simon Klaver—an application engineer from Sagantec who introduced me to all the secrets of migration tools and provided a general presentation that is on the CD.

Jim Lindauer—from Tanner Research, he agreed to provide me with a free copy of L-Edit software for the writing of the book. Special thanks to Tanner Research for providing a demonstration copy of their layout editor including the cross-sectional viewer so that the readers of this book can experience the thrill of IC layout design.

But most of all I thank Gregg Shimokura, the technical contributor to the book. We worked together in MOSAID for more than 5 years, and he was always ready to help me and others to know more about VLSI design. During this time he became the Manager of the IC CAD Technologies group, and we worked together to develop new methodologies that can enhance design capability. After so many years of wanting to write this book, I began because he offered voluntarily to help me. Everything you will read in this book was initially started by me, but Gregg is the master who placed them in the right flow, reviewed my English, and made many additions to the raw material that he had to work with. Gregg added to this book the engineering view. We hope this view will help students understand how to become better engineers by knowing more about the results of their work in layout. Thank you again, Gregg, for all the long nights and working weekends that helped this book to be born.

CHAPTER ONE

Introduction

1.1   HISTORY OF THE PROFESSION

During the past two decades, the electronics industry has grown very fast both in size and in complexity. Designers began talking about chip design only 25 years ago. At the beginning, the idea was to design chips to reduce the computer size. Instead of room-sized computers, we have now ended up with PCs running at a speed that back then was considered "impossible to imagine." The application of IC technology has exploded into many parts of our lives.

IC layout design was originally hand-drafted on special paper called Mylar. This was a long and laborious task. The market demands and advances in technology brought about an immediate need to develop software and hardware solutions to improve the time-to-market of the chip designs and especially to automate the entire process. Accuracy of the final masks was also a driving force in the computerization of layout design.

The first platforms were custom built to ensure that graphics applications ran quickly and had sufficient capabilities. Companies such as CALMA (Data General) built mainframe-sized machines and developed specialized software for printed circuit board (PCB) and integrated circuit (IC) applications.

The disk size was huge by today's standards. The top-of-the-line computer had 220 MB of disk space and only 0.5 MB of DRAM was available at the time. The price tag was around $1 million U.S., and not everybody could afford to be involved in this kind of design. As the market and the chip sizes grew and more companies were involved in chip design, the hardware and software developers came up with faster, smaller, and cheaper solutions.

The biggest revolution in hardware was the development of the "engineering workstation," which ran a version of the UNIX platform. Workstations have developed over the years to incredible speed and complexity. They are used for all kinds of engineering design, so the prices are very affordable. HP, Sun, and IBM are only a handful of survivors in this field, Daisy being one that has disappeared from the market. Today there is tremendous pressure to go to even cheaper and more popular platforms, such as PCs with Linux and Windows NT platforms.

As the hardware platforms evolved, software development progressed at an even faster rate. Companies such as Mentor Graphics, Cadence, Compass, and Daisy gained larger and larger shares of the IC and PCB design tools market. For the PC platform, a company such as Tanner, with a product called L-Edit, is an example of how the software development market has grown for IC design (more details are given in Chapter 10).

The direction for development of the software has really been toward more and more automation of the tasks that are labor intensive: for example, designs with hundreds of transistor blocks, where interconnection analysis is impossible to do by human eyes, or verification of a 256-MB memory chip (more details in Chapter 10).

Significant examples of automation include the following:

Layout synthesis: Layout can be created from "code" instead of the traditional methods of manually drawing the polygons.

Layout migration: Alternatively, layout can be "migrated" from one set of design rules to another using mapping and sophisticated compaction techniques.

Layout verification: These tools perform an increasing number of checks on the final layout before it goes to production. For example, minimum size rules are checked to ensure that the design is manufacturable.

Circuit synthesis: Similar to layout synthesis, in this case schematics can be automatically generated from specialized "code" (i.e., VHDL or Verilog). This has had a huge impact on layout design, as the sheer volume of circuitry produced by these circuit synthesis tools created a need for more layout automation such as place-and-route tools.

Place-and-route: Instance placement for literally millions of cells as well as optimizing the placement for minimum connectivity and maximum circuit performance.

Today, layout design is carried out in an environment that is ever changing. The software tools and approaches, computing platforms, the companies providing these tools, the customers we serve, the applications that are being implemented, and the market pressures we face are all changing year by year.

These changes make this industry an interesting one in which to be involved. However, let's not forget that the fundamental concepts behind producing quality layout are based on physical and electrical properties that never change. This is the basic principle on which this book was written.

1.2 WHAT IS LAYOUT DESIGN?

We define layout design as follows:

The process of creating an accurate physical representation of an engineering drawing (netlist) that conforms to constraints imposed by the manufacturing process, the design flow, and the performance requirements shown to be feasible by simulation.

Let's look at this definition in greater detail as there are numerous implications buried within.

A process: First and foremost, layout design is a process with many steps that should be followed in a logical order for optimal results. For example, the "process" of layout design may include setting up a database or suite of tools with the appropriate layers; defining the floorplan of each cell or chip; and/or running verification checks in the proper order.

Creation: "Design" and "creation" are usually synonymous, and layout design is no exception. Implementing one schematic in two different technologies usually results in layouts that look quite different, thus demonstrating the creative nature of the trade. In the same way, a schematic that will be used in two different regions of the chip may result in two different architectures, adapted to their geographical location.

Accuracy: Although layout design is a creative process, we must not forget that the first requirement of the final layout must be that it is equivalent on a transistor-by-transistor basis to the engineering drawing. Redesigning the configuration of transistors to "improve" the circuit is not the role of the layout designer unless you plan to take over (or already have taken over) the circuit design task as well.

Physical representation: CMOS ICs are made using an extremely complicated process that in the end results in tiny transistors and wires being constructed and connected on a silicon substrate. Layout design is the art of drawing these transistors and wires as they look like in silicon; thus, the layout can be thought of as the physical representation of the circuit.

Engineering drawing: This may sound a bit old-fashioned, but it is accurate. Transistor-level or gate-level schematics have historically been the primary "drawing" and in many companies they remain so. Fancier methodologies these days result in some layout designers receiving a large text-based file called a "netlist." However, in order for humans to understand a netlist, it is usually accompanied by a block-level schematic or drawing. Engineers (or equivalents) are the main providers of the drawings, but as the industry changes this may change as well.

Conform: By conforming, we mean "meeting the requirements of" and not necessarily "the smallest or best design possible." There are many trade-offs to be made in the process of design: reliability, manufacturability, flexibility, and (perhaps most importantly) time to market, to name a few. Of course, there are minimum requirements that have to be met, but to achieve the optimal design at the expense of the project schedule is not practical in today's marketplace.

Constraints imposed by the manufacturing process: These constraints include layout design rules such as the smallest width a metal track can be, but also many other manufacturability or reliability guidelines that will improve the overall quality of the layout. For example, in the case of a metal track, a wider line may improve the manufacturability of the design and thus should be used where space permits.

Constraints imposed by the design flow: These constraints include guidelines established to enable all other tools that are to be used in the design flow to be able to efficiently use the completed layout. For example, some routers like to have connections to cells on a regular pitch, while others do not care. Another example is the methodology to add text to layout so that the text can be used later for identification purposes.

Constraints imposed by the performance requirements shown to be feasible by simulation: An engineer completing a circuit design without detailed knowledge of how the circuit will be implemented in layout is required to make some assumptions. For example, the engineer designing the circuit will not know the exact area of the block without implementing the circuit in layout and so must make an educated estimate based on the information available. The total area figure may be important to know so that the maximum line length within the block is also known. This normally cannot be avoided, and the trick is to try to communicate these assumptions and thus constrain the layout accordingly. In our example the total area estimate used by the circuit designer should also be used by the layout designer as a target area, and differences from this estimate on the low or high side should be fed back to the circuit designer for resimulation.

In summary, layout design encompasses many different areas; it requires many different skills; and there are many trade-offs and decisions to be made that affect the quality of the final implementation. Great layout design requires a sound understanding of all of these issues, and we hope to cover all of them in various degrees throughout this book.

1.3   IC DESIGN FLOW

Where does layout design fit in the overall scheme of things? As defined in Section 1.2, layout design occurs once an engineering drawing is complete. Let us look at layout design in the context of an IC's complete life cycle and where it fits in the "flow."

There are many kinds of design flows based on the specific design under development. Let us consider a general conceptual flow through which all product concepts pass on their way to market (Figure 1.1).

1. First, it is normally the marketing department that defines the product to be developed.

2. The definition of the architecture or behavior of the design is the next step. Circuit design engineers decide the architecture of the chip to perform the market and/or IDEA functions.

3. System simulation is done by a group of engineers who define and verify the definition of the individual blocks to be integrated into the final chip. This step validates that the architecture defined in step 2 is sound and clearly defines manageable blocks to implement further.

4. Circuit design groups perform all the digital and analog simulations to verify the circuit solutions and gate connectivity, as well as the sizes of the gates

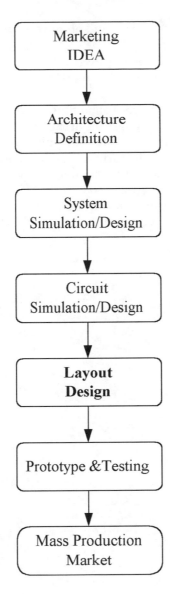

Figure 1.1 IC design flow.

(to meet timing specifications). These groups interface with the layout design groups who adapt the circuit to the floorplan of the chip.

5. Layout design is done by engineers and layout designers. Their work consists of laying out polygons. Transistors, substrate connections, connections (using 1 to 6 layers of metal), etc., are implemented for all of the blocks using the schematics generated by the circuit group. The final design going to mass production is the layout of the entire chip.

6. After the first wafers are manufactured, a group of test engineers will try to test the chips. First, they will check if the process parameters are within the acceptable tolerance levels. The following step is to test the chips using an

engineering tester in order to find all the specification violations and to try, on the spot, to fix them.

7. If and when all the errors are fixed (process and/or logical), the chip will move to mass production and to market.

Remember that this is a conceptual flow. In reality, there are many feedback loops and iterations of the design as it moves through the different stages. Changes to the design occur as a result of many different factors, including many that arise from layout limitations or constraints. Anticipating these issues or problems before they occur is where understanding the basic fundamentals differentiates great designers from good ones.

Where do we start? From a layout designer's point of view, the work starts once a schematic or netlist is created. On to Chapter 2.

CHAPTER TWO

Schematic Fundamentals

You have been given or have designed a schematic and are ready to move to layout. What's next? In this chapter we will learn the basic building blocks of a schematic and the fundamentals of preparing yourself to implement the design in layout. We start by presenting the basic building block of all CMOS circuits—the transistor. We then continue by making sense of a typical schematic drawing, and we also lay the groundwork for more advanced topics.

2.1   THE MOS TRANSISTOR: THE BASIC CIRCUIT STRUCTURE

The transistor is the smallest building block or device that we need to understand to effectively implement or layout a design. Let's first consider the functionality of the transistor and try to provide a basic understanding of the operation of a transistor so that we can maximize the performance of the design.

CMOS stands for complementary metal oxide semiconductor. This name is appropriate because there are two flavors of transistors, PMOS and NMOS, and together they complement each other, as we shall see in this section. Typically, a schematic might denote PMOS and NMOS transistors as shown in Figure 2.1. Note that the drain and source nodes are reversed as drawn in the diagrams.

In most cases the "Bulk" connection is always connected to the logical "1" level for PMOS and logical "0" level for NMOS. For this reason most schematics do

Figure 2.1   PMOS and NMOS transistors.

Figure 2.2 PMOS gate open and NMOS gate open.

not show the bulk connection; it is implied. Of course, this is not always the case. For the moment, in the following schematics we will ignore the "bulk" connection.

The gates of the PMOS and NMOS transistors are open or the transistors are "on" under different conditions. PMOS transistors are "on" when the gate is at a logical "0" level. Conversely, the NMOS transistor is "on" when the gate node is at a logical "1" level. The way to remember this is that the bubble on the gate of the PMOS looks like a "0" and the NMOS gate looks like a "1" (Figure 2.2).

Both transistors operate very much like a "switch" or a valve in a water pipe. Like a valve, the "gate" controls whether the switch is open or closed. Positive current flow is defined as the action of "draining" water or charge from the drain side of the transistor to the water or "source" side when the gate is open. If the gate is closed, current (or water) does not flow.

A simpler way to visualize the operation of the transistors is as a resistor when it is "on" (Figure 2.3).

The amount of current that flows through the transistor is limited by the equivalent resistance of the transistor. As we shall see later, the sizing of the transistors directly affects this equivalent resistance. We will use this simpler resistor model in analyzing the operation of the transistors from this point on.

Now let's consider the case when the source is connected to a static logic level. Generally, logical "1" levels are denoted on a schematic by the highest supply voltage for the design. Typically this high supply voltage would be labeled as VDD, VCC, or perhaps VPP. Conversely, logical "0" levels are denoted on a schematic by the ground level of the chip. VSS, GND, or GROUND are typical names. Under these conditions and with the gates of the transistors open the drain nodes are naturally driven to the same level as the source.

Due to the physical nature and limitations of the PMOS and NMOS devices (not to be discussed here), PMOS transistors are almost always used to establish logical "1" levels and NMOS logical "0" (Figure 2.4), although there are exceptions, of course. This is why PMOS and NMOS together have been termed "complementary": they complement each other because, together, they simply and reliably generate both logic levels. For this reason, Boolean logic is easily implemented using PMOS and NMOS transistors, which is one of the main reasons why CMOS circuitry is so popular today.

Let's not completely forget the bulk connection mentioned earlier in this section. Remember that the bulk is generally connected to the respective logic levels, and the implied connections to the supply levels are shown in Figure 2.5.

The size of the transistor should also be identified on the schematic (Figure 2.6). Each PMOS and NMOS has a length and a width. These dimensions will be

Figure 2.3 PMOS resistor model and NMOS resistor model.

Figure 2.4 PMOS generating a "1" and NMOS generating a "0."

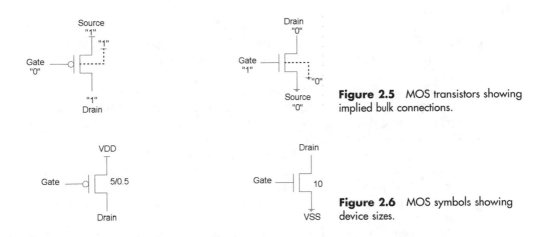

Figure 2.5 MOS transistors showing implied bulk connections.

Figure 2.6 MOS symbols showing device sizes.

explained in detail in a later chapter, and for now take this as a given. Typically the length of either transistor may not be shown and has a default value. This value is usually the minimum allowable as limited by the process technology, and it is this number that is quoted to specify the technology. For example, a 0.25-μm process typically means the default gate length is 0.25 μm and thus is not shown on the schematic because it is redundant information.

In Figure 2.6 the width of the PMOS transistor is 5 μm, and that of the NMOS is 10 μm. Generally, the width value is always stated first. The PMOS transistor length is 0.5 μm, and since the NMOS is not shown, it is assumed to be the default value for the process, which is 0.25 μm.

When we start to look at the layout of transistors, it should become more obvious that the resistance of the transistor will decrease and the current drive of the transistor will increase as the width of the transistor is increased or the length of the transistor is decreased. For this chapter, please take this as a given.

2.2   LOGIC GATES

The majority of schematics today are not filled with transistors. The reasons for this are many, but the main ones are that it is impractical because of the complexities of the designs that are undertaken, and that transistors are grouped into what is called a logic gate or "gate." A logic gate could be confused with the gate of a transistor, but we hope that the context in which the term is used will be sufficiently obvious.

Logic gates are implemented directly or in combination to form Boolean logic functions. Theoretically, almost any Boolean logic function can be implemented with a single logic gate, but in practice this is not done. We hope that, after reading this book, you will fully understand why.

In general, most logic functions are implemented in CMOS using inverters, two to four input NANDs, two to four input NORs, and transmission gates. Let's begin to learn about these gates by understanding the simplest of all logic gates: the inverter.

2.2.1 Inverter

As the name implies, the inverter is the simplest logic gate. Its function is to invert the signal received on the input node to the opposite polarity to the output node (Figure 2.7).

Let's use our knowledge of transistors. Knowing that the PMOS is "open" when receiving a "0" means that the "1" is driven to the output. In this case the NMOS is off and does not affect the output level. Conversely, by the same rules, a "0" is produced when the input is a "1" (Figure 2.8).

CMOS logic by its very nature is always inverting. Also note that the NMOS and PMOS are never "on" at the same time. This demonstrates the reason why CMOS is a low-power style of circuit design. Once the gate switches state, there is no DC current path between VDD and VSS; such a path, if it existed, would consume DC power.

In specifying the inverter size, now two device sizes are required (Figure 2.9).

- The "P" and "N" identifiers specify the device type. Again, generally the widths are stated first.
- In this case the PMOS transistor width is 2 μm, and that of the NMOS is 1 μm.
- The PMOS transistor length is 0.5 μm, and since the NMOS is not shown it is assumed to be the default value for the process.

In the next sections NAND and NOR gates will be covered. NANDs are inverted AND gates and NORs inverted ORs. They both are single-stage gates, and this is one reason why they are the basic building blocks of CMOS logic.

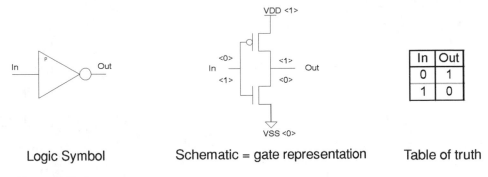

In	Out
0	1
1	0

Logic Symbol   Schematic = gate representation   Table of truth

Figure 2.7 Inverter.

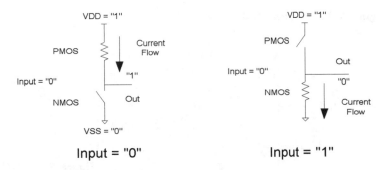

Input = "0"   Input = "1"

Figure 2.8 Inverter operation.

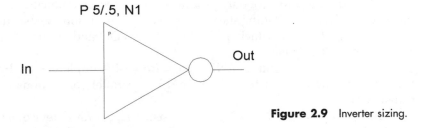

Figure 2.9 Inverter sizing.

2.2.2 Two-Input NAND Gate

When a logical decision is required to be made between different signals, NAND and NOR gates will do the job. By following the operation of the individual transistors under each input condition in the truth table of Figure 2.10, you will see that the desired output is produced with the transistor configuration shown.

The "Not AND" function (OUT = "0") is produced when both IN1 and IN2 are both "1." The requirement for both inputs to be "1" simultaneously is achieved by connecting the two NMOS transistors in series. At the same time, the PMOS transistors are connected in a complementary fashion by being in parallel.

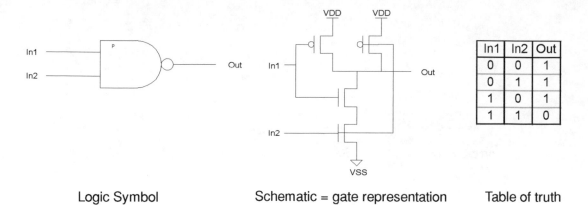

| | Logic Symbol | | Schematic = gate representation | | Table of truth |

In1	In2	Out
0	0	1
0	1	1
1	0	1
1	1	0

Figure 2.10   Two-input NAND gate.

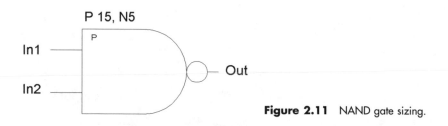

Figure 2.11   NAND gate sizing.

This configuration not only produces the correct functionality from the gate, but also results in eliminating static DC power consumption by ensuring that there is never a condition in which a PMOS path to VDD and an NMOS path to VSS are "on" simultaneously.

Three or more input NAND gates are easily implemented by extending the series connections of the NMOS and the parallel connections of the PMOS transistors.

In specifying the NAND gate transistor sizes, four device sizes are now required. In most cases, however, all PMOS transistors will be the same size and, similarly, all NMOS transistors will be the same size; therefore, once again typically only two values are required (Figure 2.11). This is also true of NOR gates, and indicating sizes on the NOR gate is done in a very similar way.

- The "P" and "N" identifiers specify the device type. Again, generally the widths are stated first.
- In this case the PMOS transistor width is 15 μm, and 5 μm for the NMOS.
- The PMOS and NMOS transistor are assumed to be the default value for the process.

If distinct sizing for the two separate PMOS transistors is required, typically this would be indicated by a subscript to the "P" identifier such as "P1, P2," and additional values would be given.

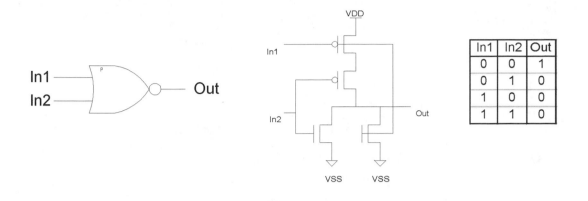

| Logic Symbol | Schematic = gate representation | Table of truth |

Figure 2.12 Two-input NOR gate.

2.2.3 Two-Input NOR Gate

The NOR gate is the mirror or complementary configuration to the NAND. In the NOR gate the series/parallel connections are reversed between the NMOS and PMOS transistors—the PMOS transistors are in series and the NMOS in parallel (Figure 2.12).

Once again, the potential for DC power consumption is eliminated under all input conditions, and three or more input variations of the NOR are easily made by increasing the series and parallel connections of the PMOS and NMOS transistors, respectively.

Transistor size values are indicated in much the same way as for NAND gates, and a description of a typical convention will not be repeated here.

2.2.4 Complex Gates

As mentioned previously, almost any Boolean logic function can be implemented in a single-stage CMOS logic gate. The term complex gates is the name given to logic gates that have a "complex" function, usually a combination of AND, OR, NAND, and NOR, all implemented in one logic stage.

Because complex gates are implemented in a single stage, in almost all cases power consumption, area, and speed benefits are achieved.

Figure 2.13 is an example of a complex logic function implemented in multiple gates.

If we do a simple transistor count for this logic we find that there are 16 transistors in all with 3 stages of logic. It is very common to find that an engineering schematic would not be designed this way but in a single stage of logic represented by a symbol such as that shown in Figure 2.14.

By combining the inverters with their respective driving gates, you can see that the NAND–inverter combination becomes an AND and the NOR–inverter combination becomes the OR. The output NOR remains the same.

What does the transistor representation of this gate look like? We need this representation to do our layout design.

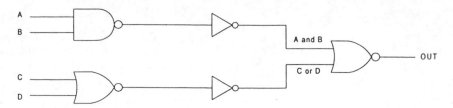

Figure 2.13   Complex logic.

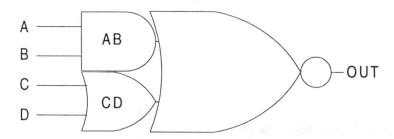

Figure 2.14   Complex gate example.

This type of complex gate is very efficient to use and build, but somehow cumbersome to draw. To determine the transistor representation we analyze the logic starting from the output gate and work backward (i.e., from right to left).

First consider the output of a two-input NOR. The idea is to combine a NAND function representing the AND gate as well as a NOR function representing the OR gate into the output NOR to create the final logic gate.

Why do we use an input NAND instead of AND? Similarly, why NOR instead of OR?

The answer is that the output NOR gate provides an extra stage of logic inversion, which we take advantage of in implementing the final gate. Since there is an inherent inversion in the output NOR gate, we do not need to implement input AND or OR functions; NAND and NOR functions are just what we need. It is wise to work this through and prove it to yourself.

Before we can perform the transistor merging as described later, the preparation step is to determine the logic gates at the input that will be merged into the output gate. This is done by simply inverting the logic at the inputs. In our case we invert the AND to NAND and the OR to a NOR.

1. We replace the AB PMOS transistors with the parallel PMOS transistors of an input NAND and the AB NMOS transistors with the respective series NMOS transistors of the same input NAND.
2. Now we use the same methodology, but for the CD devices. Replace the CD PMOS transistors with the series PMOS transistors of the input NOR and the CD NMOS transistors with the respective parallel NMOS transistors. There—you're done! (See Figure 2.15.)

If you check the truth table of the final configuration you should find that the 8-transistor logic gate is logically equivalent to the 16-transistor, 3-stage logic function presented earlier.

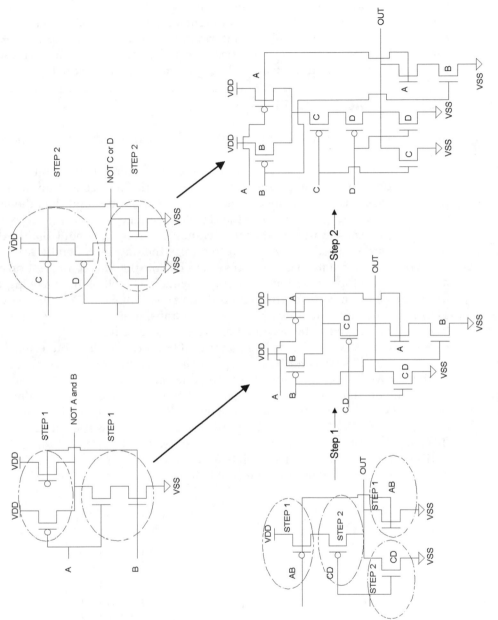

Figure 2.15 Complex gate solution.

Use this technique to expand and understand the simplicities of complex gates!

Because of the greater number of transistors for a typical complex gate, individual transistor sizes may or may not be indicated on the schematic. In most cases each transistor would have a different size, and so transistor sizes are typically omitted from the symbol. Size information must be determined by looking at the transistor-level schematic. Even if sizes are indicated, the mapping of these sizes to the transistor configuration should be manually checked before layout begins.

2.3   TRANSMISSION GATES

Let us consider one more configuration of transistors that may appear in a schematic.

In the case of the inverter, the source of both transistors is connected to a power supply. In the case of combination gates, series connected transistors form part of a chain that eventually connects to a power supply, and thus the transistors should be treated similarly to the simple inverter.

The transmission gate is a fairly common case where both the drain and source nodes are used as signals. In this case, the output generally follows the input based on the state of the controls A and B. Note that this configuration allows for noninverting propagation of the input signal, as well as the blocking of the input signal when both control signals disable the PMOS and NMOS transistors. These are powerful features of this gate; transmission gates are used quite frequently and need to be designed carefully (Figure 2.16).

Remember we said that in general PMOS transistors are connected to generate logical "1" levels and NMOS logical "0," and almost never the reverse. The truth table for the transmission gate shows one of the reasons why this is so. PMOS transistors are able to pass "0" levels, but they do so somewhat unwillingly and degrade the "0" level. The same is true for NMOS transistors and "1" levels. This is what is meant by "Weak Levels" in the truth table. Unless specifically intended, these weak-level conditions are generally avoided in robust logic designs. Usually both controls are implemented such that the transmission gate is either completely "on" or "off" (both transistors) but not halfway.

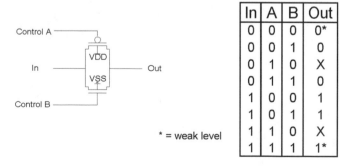

In	A	B	Out
0	0	0	0*
0	0	1	0
0	1	0	X
0	1	1	0
1	0	0	1
1	0	1	1
1	1	0	X
1	1	1	1*

* = weak level

Schematic = gate representation   Table of truth

Figure 2.16 Transmission gate.

2.4 UNDERSTANDING THE SCHEMATIC CONNECTIVITY

In implementing the layout of any schematic, there is more to the final design than is explicitly shown. Connections appear on a schematic as a simple line drawn from point A to point B, or a simple connection of two transistors in series or in parallel. In reality, a line represents a signal path that needs to be physically implemented and optimized. Let's look at an example (Figure 2.17).

The gates and transistors should look familiar, and the different transistor representations of the various gates have been described. The challenge now is to understand the connectivity of the devices. We have already seen that a bulk connection to each transistor is required but is not explicitly shown. Table 2.1 outlines the different types of symbols.

TABLE 2.1 Schematic Connections

Schematic Representation	Description
———	Simple wire connection. These signals are local signals to be routed and implemented within the schematic under consideration.
≫	On page connector. A virtual connection is achieved with this symbol. The connection name or node name is used to identify where on the schematic the net is to be routed. In our example the two nodes labeled CLKD are electrically connected but are not visibly connected. Generally, this is done to avoid cluttering the schematic with wires.
⊏▷	Port or pin connector. This symbol identifies a net that enters or exits the schematic under consideration and is part of the "interface" of the schematic to the outside world. These signals may have special considerations attached to them for performance or reliability reasons, so it is important to find out if such conditions exist.
VDD	Global connector. We have seen this as the bulk connection to the transistor. A global connector identifies an electrical node that is required internally and externally to the schematic block. The "VDD" net in this case is used everywhere and is global. Again, drawing the wires to show the implied connectivity is impractical.

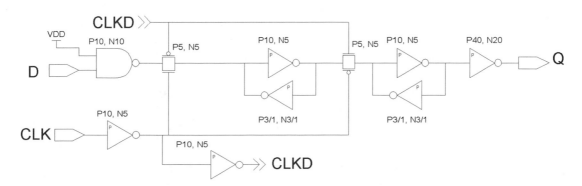

Figure 2.17 Schematic example.

2.5   REVIEW OF FUNDAMENTAL ELECTRICAL LAWS

IC layout design is fundamentally the art of implementing an electrical circuit in terms of polygons and shapes, which represent transistors and connections to form the final design. The important concept that we must not forget is that the final design will have electrical characteristics that are very much defined by the characteristics of the physical layout.

The intent of this section is to review a few basic electrical laws and principles that should be understood, so we can establish a good foundation upon which we can move forward and develop efficient and effective layout methodologies.

2.5.1 Ohm's Law

This is the most basic and fundamental law:

$$V = I \times R$$

Voltage = Current × Resistance

We have seen that MOS transistors operate as "resistors" when they are "on" or when the gate is "open." The current flow induced by the opening of the gate creates a voltage swing across the transistor. This demonstrates the application of Ohm's law! Given the resistance of the transistor and a positive current value, the resulting voltage change is explained by Ohm's law (Figure 2.18).

Similarly, when the gate is "off" the current is "0." By Ohm's law, the voltage change is also "0," which makes sense since the gate is "closed" and it acts like an open circuit.

In reality, the resistance of the transistor is dynamic, as is the amount of current flowing through the transistor. Therefore, this is a very simplistic model,

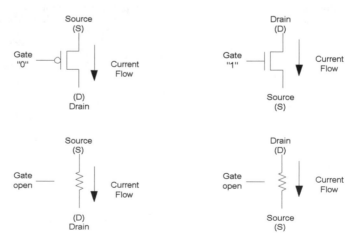

Figure 2.18   PMOS model (left) and NMOS model (right).

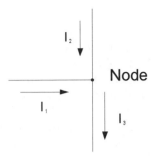

Figure 2.19 Node currents and Kirchoff's law.

but it effectively explains how Ohm's law works and gives us the concepts behind how a transistor operates.

Ohm's law is a powerful principle to remember and is the foundation for circuit and layout design alike.

2.5.2 Kirchoff's Current Law

Kirchoff's current law is another fundamental law that helps us to explain certain concepts in future chapters. Kirchoff's current law states that the sum of currents into any electrical node is to zero. In this case currents coming into a node are deemed to be positive currents by convention, and currents passing out of a node are deemed to be negative currents, so their overall sum should equal zero:

$$I_1 + I_2 + I_3 + \ldots + I_N = 0$$

Another way of stating the same thing is that the sum of currents into a node must equal the sum of currents out of a node (Figure 2.19).

2.5.3 Resistance

We have already mentioned the concept of resistance without really explaining it in more detail. We have used the resistor to model the transistor in the "on" state.

In simple terms resistance can be thought of as the inability (or ability) of a conductor to conduct charge. Using a water analogy, a pipe of large diameter has a lower resistance than a smaller diameter pipe because it can pass a larger amount of water. The cross-sectional area of the pipe is larger in this case. This assumes that the two pipes are the same lengths. As a pipe or conductor increases in length, the resistance also increases.

The convention in IC design for resistance calculation is to characterize each conductor layer in terms of resistance per "square." One "square" is defined as the condition when the length of the conductor equals the width.

The formula for calculating the resistance of a conductor is

$$R = \rho \times l/w$$

where "ρ" is the resistivity of the layer measured in $\Omega/\square$, l is the length, and w is the width of the conductor.

2.5.4 Capacitance

In simple terms, capacitance can be thought of as the amount of charge a body or conductor can hold per unit of voltage between the node in question and another reference node. Using our water analogy, a capacitor should be thought of as a dammed lake that is filled with or emptied of water based on the electrical power needs of consumers.

The amount of capacitance a conductor has is determined by the area of the conductor and how far it is away from the reference node. Again using our water analogy, let's consider a lake. How much water will it take to fill the lake (think how much charge will it take to charge up the capacitor)? The answer is, it depends on the surface area of the lake and how deep it is.

The tricky part of this concept is that the distance between the reference node, the bottom of the lake, and the surface of the lake determines the depth of the lake. *The farther the reference node is away from the conductor, the shallower the lake is. If the reference node is very close, the lake will be deeper and thus the overall capacitance is greater.* The concept behind this is that the charge in the conductor is attracted to the reference node by an electric field attraction associated with opposite charges. Closer bodies have larger electric fields and thus larger capacitance values.

There is also a dependency on the material that separates the two nodes. Some materials isolate the attraction to a better degree than others do.

A very simple model for the capacitance of a conductor is calculated as

$$C = \varepsilon \times A/d$$

where A is the surface area of the specific conductor, d is the physical distance between the conductor and the reference node, and ε is a constant representing the characteristics of the insulating layer between the conductor and the reference node.

2.5.5 Delay Calculation

Without going into gory theoretical detail, let us consider a simple example of a inverter driving a wire or conductor. The wire is represented as a single resistor and a lumped capacitance (Figure 2.20).

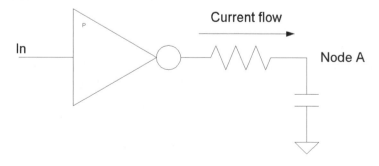

Figure 2.20   Delay calculation circuit.

Our goal is to calculate the delay from IN to node A. The total delay is dependent on two factors:

- The associated switching delay of the inverter. This inverter delay is dependent on the size of the resistor and the capacitor. This delay is normally calculated or measured from simulation, so we will not consider it formally here.
- The delay of the wire is due to the resistor and the capacitor. A first-order approximation of the delay through the wire as an independent component is

$$Delay = R \times C$$

This simple equation gives us an easy formula to analyze the delay through different wiring scenarios and allows us to make the appropriate trade-offs in laying out the final design.

If it is required to minimize the delay through a given circuit, we need to consider reducing both the resistance and the capacitance of the wire. Using our knowledge of resistance and capacitance, we can optimize our layout to minimize the delay by doing the following:

- Minimizing the length of the conductor. This reduces both the resistance and capacitance terms.
- Optimizing the width of the conductor. Decreasing the width of the conductor decreases the capacitance of the wire; however it increases the resistance!
- Increasing the spacing of the conductor to other reference nodes. This decreases the capacitance of the wire. Usually this means running the wire in areas that are free from other polygons or shapes or using a top metal layer instead of the lower one.

CHAPTER THREE

Layout Design

In Chapter 1 we defined in great detail layout design as follows:

> The process of creating an accurate physical representation of an engineering drawing that conforms to constraints imposed by the manufacturing process, the design flow, and the performance requirements shown to be feasible by simulation.

Summarizing once again, a layout designer is a person who knows basic electrical concepts, process limitations, and properties; has a talent for seeing and feeling space and floor plans; and can learn and use various CAD tools.

Let us understand in greater detail the manufacturing process and how it relates layout to the physical representation of the design.

3.1 INTRODUCTION TO CMOS VLSI MANUFACTURING PROCESSES

There are many kinds of design processes, but this text discusses only CMOS technologies. We will first discuss the manufacturing order of layers (Figure 3.1) without going into the details of how each step is physically realized.

We start with a bare silicon wafer. Between steps an isolation layer is grown to protect areas that are not to be patterned.

P and N bulk regions are defined by differentiating different areas of the wafer with "wells" or "tubs" of the appropriate type.

The polysilicon that forms the gate areas is added next.

Source and drain areas are defined by diffusing areas on either side of the gate polysilicon. Other active areas such as substrate contacts and guard rings are formed at the same time.

In order for interconnect layers to be connected to the polysilicon and/or active areas, contact holes are created in the isolation layer on top of the layer to be connected.

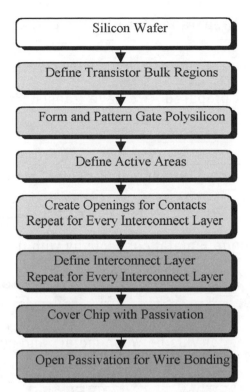

Figure 3.1 CMOS manufacturing process.

The interconnect layers are deposited and fill the contact holes created in the previous step.

The last layer is called the passivation layer with openings for wire bonding connections. The passivation layer is a glass layer that isolates the chip from the external world.

This diagram is a very simple explanation of the manufacturing process. Different process technologies have significantly different manufacturing steps.

DRAM memories for example have four layers of polysilicon to construct the memory cell capacitor. ASIC designs have only one polysilicon and more layers of metal, which are used to connect many, many logic gates. Using five to six layers of metal, microprocessors, and other complex ASIC designs can be produced (Figure 3.2).

3.2   LAYERS AND CONNECTIVITY

Let us simplify the types of layers that are used and introduce the concept of *mask* layers and *drawn* layers.

If we analyze most CMOS processes, we find that there are four basic layer types:

1. *Conductors:* These layers are conducting layers in that they are capable of carrying signal voltages. Diffusion areas, metal and polysilicon layers, and well layers fall into this category.

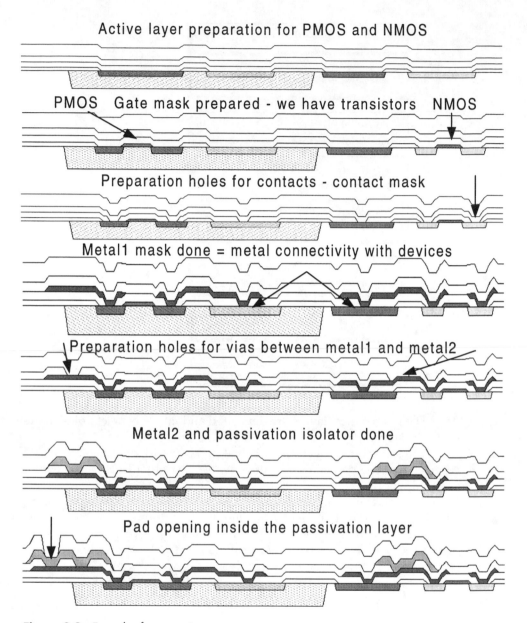

Figure 3.2 Example of cross-section process steps.

2. *Isolation layers:* These layers are the insulator layers that isolate each conductor layer from each other in vertical and horizontal directions. This isolation is required in both the vertical and horizontal direction to avoid "short circuits" between separate electrical nodes.

3. *Contacts or vias:* These layers define cuts in the insulation layer that separates conducting layers and allow the upper layer to contact down through the cut or "contact" hole. Metal vias or contacts are examples of these. Openings

in the passivation layer for bonding pads are another example of a contact layer.

4. *Implant layers:* These layers do not explicitly define a new layer or contact, but customize or change existing conductor propriety. For example, diffusion or active areas for PMOS and NMOS transistors are defined simultaneously. A P+ mask is used to create P+ implant areas that define certain diffusion areas to P-type by the use of a P-type implant.

Using a combination of these four types of layers, transistor devices, resistors, capacitors, and interconnections are created.

In almost all cases, the number of layers that are drawn by the layout designer has been reduced to the minimum number required for the mask-making process. This minimum number of layers is referred to as the set of *drawn* layers. Minimizing the number of drawn layers reduces human error and layer management, as well as the computational requirements of the CAD software.

The *mask* layers, or the layer shapes that are translated to the optical masks, are sometimes different from the drawn layers. First, there may be many more mask than drawn layers. In this case, the additional mask layers are automatically generated from the drawn layers.

Additionally, the mask layers may be resized from the drawn layers to account for variances in the manufacturing process. This resizing is also done automatically by the mask-making process.

Note that isolation layers are never drawn but are always implied from the mask layers as part of the manufacturing process.

From this point on any reference to a layer should be interpreted as meaning a drawn layer.

Of course, all of the layer entry is done with sophisticated CAD software, and the subsequent manipulation of layers is also done with computers and complicated software.

Every shape that is drawn is entered either as a "polygon" or a "path." There are subtle differences between the two, which are partly related to the way computers handle and process the layout database. There are situations where polygons are better suited to layout than paths, and vice versa. These differences will be explained in the next two sections.

3.2.1 The Polygon

As the name implies, a polygon is an N-sided shape that geometrically has $N + 1$ vertices, which define the shape (the computer sees $N + 1$ vertices because there is one vertex that is double counted because it is counted as both the origin and the end point).

The typical uses for polygons are places where the designer has to cover areas that are not necessary a simple rectangle—for example, cell boundaries, transistors, n-wells, contacts, diffusion areas, and transistor gates. In addition, polygons are flexible enough to be used to define areas because they can be implemented in various angle modes such as 90 or 45 degrees or in some rare cases as freehand shapes.

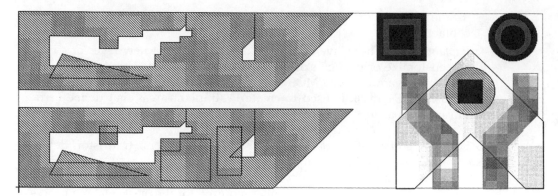

Figure 3.3 Examples of polygons.

The pros of using polygons include the following:

- Can be used to enclose an odd-shaped area
- Can be easily drawn, added to, or subtracted from
- Can be easily merged with other polygons at the same level of hierarchy and same layer

See some examples of polygons in Figure 3.3.

The cons of using polygons include the following:

- Not easy to modify complex polygons for consistency. An example might be when a uniform width is desired and to modify all portions of a polygon is tedious.
- Requires more computer database space compared to a "path" in situations where paths are useable.

3.2.2 The Path

As the name implies, a path is a shape that is defined by a start and end point, intermediate vertices, and a width value. It is used primarily to connect devices and run signals from point to point because a path has a *consistent width*.

A path is easily manipulated and uses fewer computer resources than a polygon in terms of data. The vertices define a centerline (or sideline) for the path, and an additional variable defines the path's width. Path lines can also follow 90-degree, 45-degree, or freehand angle modes.

Paths can be designed as centered, left, or right justified. This means that the shape of the path appears either centered or to the left or right sides of the vertices.

An additional attribute of a path is the way the path is ended. The length of the path relative to the start and end points can be fixed, extended beyond the end points by a certain amount, or perhaps rounded.

All of these features need to be implemented with many things in mind: the target manufacturing process, the CAD tools, and design requirements. Some examples of paths are shown in Figure 3.4.

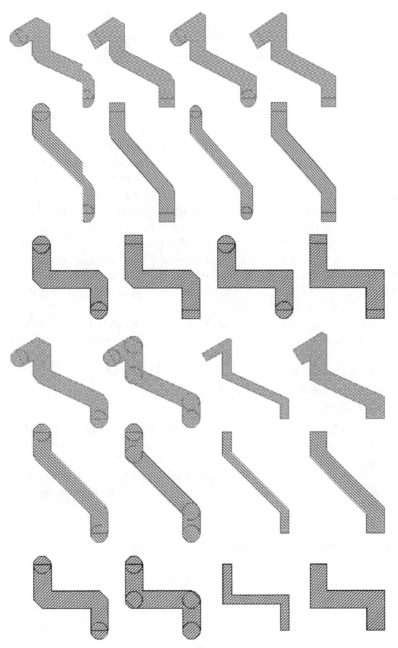

Figure 3.4 Examples of paths.

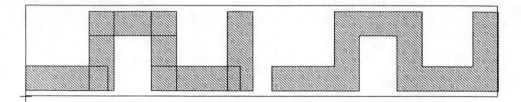

Figure 3.5 Multiple paths flattened to a single polygon.

As we can see, the path has a lot of potential in having different termina-tions and vertex formats for different layout styles and design requirements.

An efficient use of paths is to generate layout using multiple paths. Once the desired shape is defined, we can flatten the paths to get polygons (see Figure 3.5). Generating the first version of the layout using paths is much quicker and more efficient. We can still convert the paths to polygons if so desired. The reverse is very limited. A path cannot be generated easily from a polygon.

Depending on the type of layout and the designer's working habits, the more paths that are used, the more efficient the layout is. Paths are easier to change and contain less computer data. For example, moving a section of a path requires moving one edge. Moving an equivalent section of a polygon requires moving the two edges on each side of the polygon.

Efficient work habits will save time and money in the long run because minimizing the size of the layout database minimizes several other factors:

- Disk space required to store the layout database
- Workstation memory usage while working
- Screen redraw time
- Workstation CPU time required to process the entire layout database for the mask making process

The only disadvantage of the path is that some CAD tools do not support the merging of a line of paths into one polygon when merging is desired.

3.3 INTRODUCTION TO TRANSISTOR LAYOUT

Before we start to discuss the layout of transistors, let us review the schematic fundamentals presented in Chapter 2. The top half of Figure 3.6 shows the basic symbol representations of both PMOS and NMOS transistors. The length and width of the transistors are shown. Also remember that the bulk connection is there, but is hidden from view to avoid cluttering the schematic.

In Chapter 2 we stated that the amount of current flow is determined by the device size. We hinted that the current flow is increased as the width of the device is increased or the length of the device is decreased. Let's see why this is by under-standing the physical characteristics of the transistor as determined by the layout of the device.

Figure 3.7 shows a simple MOS transistor layout. Note the following:

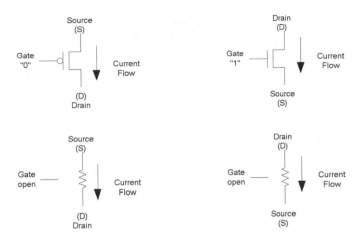

Figure 3.6 PMOS and NMOS transistors.

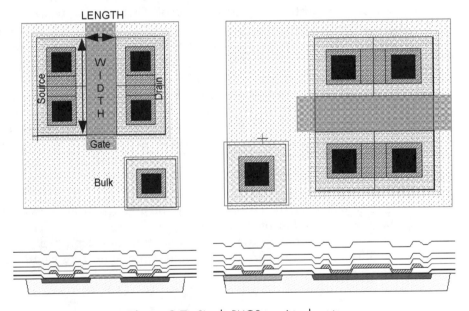

Figure 3.7 Simple PMOS transistor layout.

- All four terminals of the transistor are shown and labeled.
- The gate of the transistor is defined by a polygon of polysilicon.
- Areas of active or diffusion adjacent to the gate of the transistor define the source and drain areas. Note that the source and drain labels are in fact interchangeable!
- This transistor happens to be a PMOS transistor and the active areas are doped P-type by the P+ implant layer.
- This PMOS transistor is located in an N-type well called an N-well. This forms the transistor bulk node.

- An N-type active area (without the P+ implant layer) forms a connection to the N-well because the N-well and active areas are of the same type (N-type).
- The source, drain, and well connection are themselves connected by another contact layer. This contact layer would typically be the contact layer for the first layer of metal.
- The width and length are labeled correctly. The width is greater than the length!

The length and width of a transistor are the two most important dimensions of a transistor that we need to fully understand.

As we stated previously, when people in the industry talk about the gate size of a specific technology, they are referring to the minimum gate length. Note the following:

- In terms of layout design, the length of the transistor is the distance between the source and the drain of a transistor. This may not be intuitive, because the physical dimension of the transistor length is smaller than the width. The next paragraph should explain the reasoning behind this convention.
- In terms of transistor performance, the length of the transistor is the distance electrons have to travel when the gate is "on" or "open" to produce a measurable current flow. Remember, it is the gate voltage that controls the flow of current. If the distance between the source and drain is reduced, the gate voltage has a stronger influence in enabling current flow. The bottom line is that in the same process technology, if two transistors have the same width but different lengths, the transistor with the shorter gate length will produce more current. More current conceptually means faster performance.
- The length of a transistor in terms of manufacturing capabilities is the narrowest possible piece of polysilicon (poly) that can be manufactured reliably. Smaller poly dimensions and thus smaller transistors results in smaller ICs, so it is attractive to use the minimum gate length to minimize chip area.

Let's now consider the width of a transistor.

The width of a transistor should be thought of as the number of parallel channels that are available for current to pass from the source to the drain. Wider transistors have more channels available; more channels mean more current.

Once again comparing two transistors, this time each having identical gate lengths but different gate widths, the transistor with the larger gate width will produce more current.

To help you remember the convention of transistor length and width, think of a transistor like a bridge. The length of the bridge is the distance between the two sides of the river and the width of the bridge is the number of lanes of traffic that the bridge can accommodate. The amount of traffic that can cross the bridge is limited by the length and width of the bridge in the same way that current is limited by the length and width of the transistor.

If the design of the bridge is to allow 100 cars to cross over in 1 minute, then the bridge needs to be made wide enough to achieve this goal. In most cases the length of the bridge is fixed (similar to the minimum allowable gate length) and the only degree of freedom we have to achieve our goal is to adjust the width.

One last concept to consider. There are cases when we might want a slow or weak transistor! This is easily achieved by minimizing the width of the transistor and/or increasing the transistor gate length. Delay elements or weak feedback devices are examples where slow transistors are desired. It may turn out that in these cases the gate length does turn out to be greater than the width (Figure 3.8).

The first important thing to remember is the difference between the *length* and *width* of a transistor and how to apply this to transistor layout!

For completeness, Figure 3.9 shows the layout of an NMOS transistor.

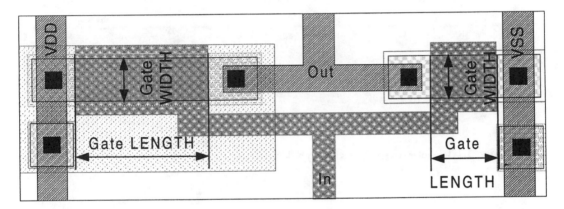

Figure 3.8 Weak feedback inverter.

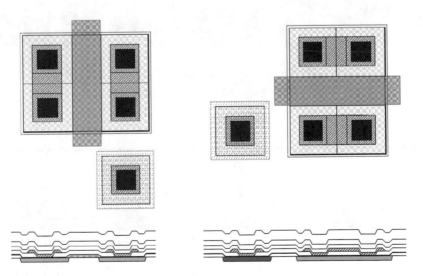

Figure 3.9 Simple NMOS transistor layout.

- This NMOS transistor is not located in any type of well and thus sits in the bare substrate. In this case, the substrate can be deduced to be P-type. The substrate forms the transistor bulk node.
- A P-type active area (with the P+ implant layer) forms a connection to the substrate because the substrate and active areas are of the same type (P-type).

3.3.1 Bulk Connections

Now that we know how the two basic kinds of transistors work and look, let's review the bulk connection node and see how it is connected. This is most easily understood by understanding a cross-section of the wafer and transistor. Note

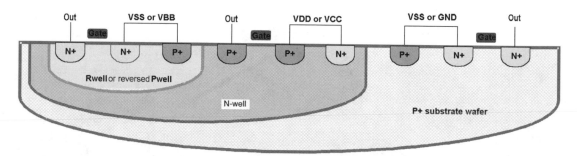

Figure 3.10 Wafer cross-section showing bulk connections.

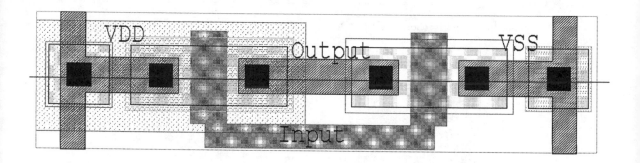

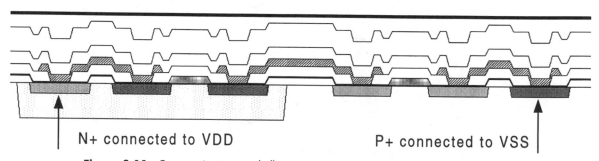

Figure 3.11 Cross-section inverter bulk connections.

that a layout designer can only understand this concept and cannot influence its design.

Most (but not all!) raw silicon wafers these days are P-type, so of the two transistor types, an NMOS transistor is the easier to design. The transistor layout is simply implemented in the bare substrate (see right-hand side of Figure 3.10).

To generate PMOS transistors we need to create a separate bulk node and therefore need another layer. This is typically called an N-well; when implemented, it forms an island of N-type substrate. Implementing P-type active regions within this N-well creates a PMOS transistor with a bulk connection as defined by the N-well (see center area of Figure 3.10).

The left-hand side of Figure 3.10 also shows an NMOS transistor design that has a different bulk node than that of the substrate. A retrograde well (R-well) or P-type well (P-well) has been implemented in the N-well. This region creates a separate P-type bulk node for the transistors implemented within this region. This is an example of substrate connection in a DRAM process.

In the case of an N-type wafer, the polarities of the transistor connections are simply the reverse of those shown previously. Figure 3.11 shows substrate connections for an inverter in an ASIC process.

3.3.2 Conductors and Contacts

From a layout design point of view, conductors and contacts are straightforward. Let's look at the formation of contacts from a manufacturing point of view so that as layout designers we can understand their use and limitations.

Different technologies have drastically different process definitions. A typical ASIC process has one type of polysilicon for the gate and two to four types of layers of metal for interconnection. An advanced ASIC process can have up to six layers of metal for interconnect and use a low-level metal called metal0 for source/drain connections. For DRAM memories, a typical process today has four types of polysilicon and three to five metals for interconnectivity. In any of these cases the conductor layer definition for the process is quite complex.

There is a subtle difference in the industry between the names *contact* and *via*. A contact typically refers to the lowest level metal hole that contacts from the lowest level of metal to the polysilicon or diffusion layers. The holes that allow higher layers of metal to connect between each other (e.g., metal1 to metal2 or metal2 to metal3) are called "vias" or "through holes."

We will use *vias* throughout this text and for easier understanding of these holes. For an illustration of contacts and vias, please refer to Figure 3.12.

As you can see from the cross-section shown in Figure 3.12, there are various isolators between the various conductor layers. I1 is the isolator between the diffusion regions and polysilicon. I2 is the isolator between the diffusion regions and metal1. A hole in this isolator generates a "contact" between the passing metal1 and the lower active source/drain layer. I3 is the isolator between metal1 and metal2, and a hole in it represents a via.

In most cases there is a distance to respect between the "contact" hole and the "via" hole, but in most modern processes the via can be placed on top of the contact. In some very complicated processes where the size of the chip is very important (read cost), the process may allow all the vias to be aligned one on top of each other. They are called "stacked" via processes.

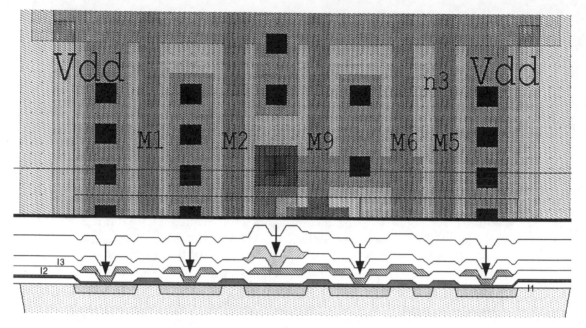

Figure 3.12 Illustration of contacts and vias.

Each metal has various characteristics in terms of resistance (R), capacitance (C), and topology requirements. Something to think about is that the higher metal layers in the process require more vias to connect down to the transistor layers. These vias add resistance. We will analyze later in the book how to deal with these electrical characteristics of the process and how to take advantage of them.

3.3.3 Inverter Layout

Now that we have all the basic concepts of transistor layout design, let's once again look at the simplest combination of transistors, the inverter (Figure 3.13).

As you can see, the transistor representation is very simple, and now we are able to generate a layout since we know how a transistor looks and where it is connected. Let's see what we can observe from analyzing Figure 3.14.

- The PMOS is connected to VDD in the schematic as well as in layout.
- The NMOS is connected to VSS in both pictures.
- NMOS and PMOS transistors have the same IN signal on their gates and same OUT on their drains—in both pictures.
- The widths are different—the PMOS is twice as big as the NMOS transistor in this example.
- The lengths look similar, but they are different, and the difference cannot be seen.
- For N-well there is a N+ connection to VDD. This connection is implied in the schematic.

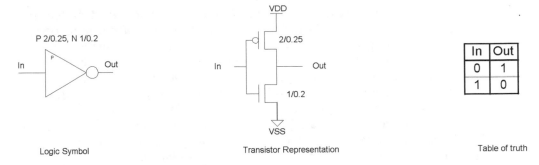

P 2/0.25, N 1/0.2

In ────────▷▷──── Out

Logic Symbol

VDD

2/0.25

In ───────── Out

1/0.2

VSS

Transistor Representation

In	Out
0	1
1	0

Table of truth

Figure 3.13 Inverter representations.

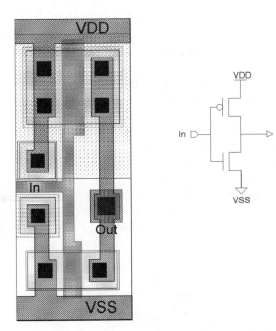

VDD

In ▷ ───────── ▷ Out

VSS

Figure 3.14 Inverter layout and transistor schematic.

- For substrate there is a P+ connection to VSS. This connection is implied in the schematic.

The design of this inverter will be presented in later sections, and this inverter is shown to give us an idea of how a complete layout cell should look.

3.4   PROCESS DESIGN RULES

Design rules are the rules that have to be respected when a given design is laid out. There are design rules for all of the components we have been introduced to: polygons and paths, transistors, and contacts. Fundamentally, these design rules represent the physical limits of the manufacturing process.

Within a company that has the capability to manufacture integrated circuits, there is a group of people who define and optimize the manufacturing process. This "processing" group defines the design rules by trading off the cost-to-manufacture and yield, among other things, against the minimum feature size that is manufacturable by the equipment and processing steps. Other factors that influence the definition of design rules could be the maturity of the manufacturing tools and process or the market requirements for an IC or foundry service.

Overall, design rules are put in place to help layout designers understand and account for physical three-dimensional limitations and manufacturing tolerances within the CAD and layout tool environment.

3.4.1 Width Rule

The minimum width of a polygon (during mask-making, all paths are converted to polygons) is a *critical dimension*, which defines the limits of the manufacturing process (Figure 3.15). The minimum gate length of a transistor is the prime example of this rule.

A violation in a minimum width rule potentially results in an open circuit in the offending layer. The manufacturing process will not reliably produce a continuous connection or wire below a specific value, and breaks in the path would result at the point at which the width rule was violated.

In addition to single polygons, width rules can also be applied to structures such as transistors or to single polygons with electrical or other special characteristics. An example of a polygon with special electrical characteristics is a metal layer that is connected to a power supply. The larger currents that pass through these metal polygons necessitate that they have a width greater than the minimum design rule, and the correct value may depend on the size of the current rather

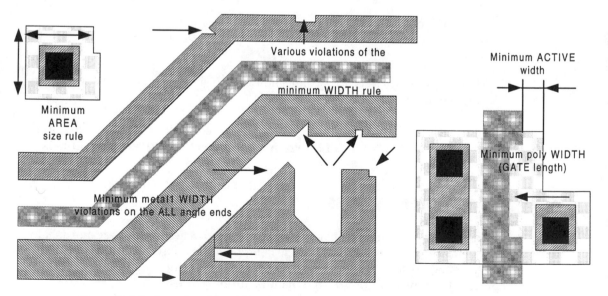

Figure 3.15 Examples of the width rule.

than being a fixed value. Large currents passing through a narrow metal track cause the track to act like a fuse, and over time or during a large current peak the metal polygon will break under the stress.

The length of a polygon (or path) is usually unlimited; however, in some processes there may be rules about minimum area requirements (for example, in the case of a contact or via where a width and a length rule together must be met). Please refer to the examples in Figure 3.15 for clarification.

3.4.2 Space Rule

Another critical dimension is the *space rule*, which is the minimum distance between two polygons. Generally, the space rule is applied to avoid an unwanted short circuit between the two polygons.

Together with the width rule on a single layer, the space and width rules define a layer *pitch*. The pitch of a layer is important when considering interconnect and routing porosity. The routing area consumed by n metal lines is easily calculated by multiplying the number of lines by the layer pitch. Please refer to the CD-ROM data for examples of pitch calculations.

Figure 3.16 illustrates the following points:

- 1 and 2 are examples of the *metal1 to metal1* minimum space rule checked in parallel and diagonally between corners.
- 3 is an example of the *poly to poly* space rule where the polygons are running in parallel at a 45-degree angle.
- 4, 5, and 6 are spacing rule examples related to *metal2 to metal2* spacing for polygons at a 90- and 45-degree angle.
- 7 is an example of the *active to active* spacing rule checked with a single distance (top example) or within a corner (bottom example).

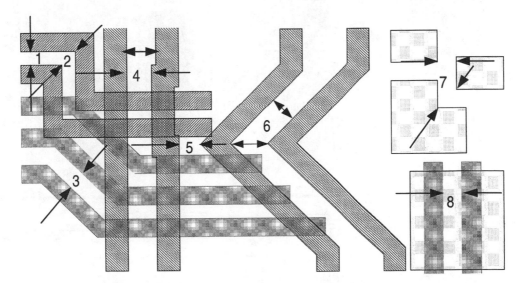

Figure 3.16 Examples of the space rule.

- 8 is an exception to example 3—the spacing rule between two polysilicon polygons may depend on their location. A typical example of this is in the case where gate polysilicon within a transistor structure has a different spacing value than that for polysilicon outside transistor structures.

Like the width rule, spacing rules are applied to polygons on the same layer, but also to polygons or structures on different layers or under different conditions. An example of a spacing rule on different layers is the spacing required between a contact to active and gate polysilicon. An example of a spacing rule between different structures would be the distance between the exposed pad circuitry and sensitive internal circuitry to ensure reliable and consistent operation under all conditions.

Many of the spacing rules defined in a set of design rules can easily be understood when looking at a process in a cross-sectional view. This is explained fully in Section 3.5, Vertical Connection Diagram.

In Figure 3.17 we can observe that the spacing between the gate polysilicon and the contacts is not the same in the two transistors. In looking at the cross-sectional view, the first thing to note is that the source and drain areas of the two transistors are not the same.

More importantly, the spacing rule of the contact to the gate polysilicon on the left-hand transistor has been violated to the extent that the gate polysilicon has been placed directly underneath the contact. A short circuit between the metal and the gate polysilicon has been created. We can easily observe the problem in the three-dimensional view. The cross-section cut line was placed in the middle of the lower contacts.

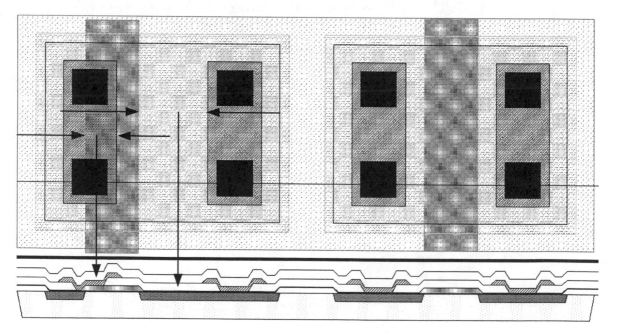

Figure 3.17   Another example of the space rule.

3.4.3 Overlap Rule

As its name implies, the *overlap rule* is defined as the minimum overlap or surround of one polygon by another. The overlap of a metal layer over a via or contact is a prime example of this rule.

Note that this rule always involves polygons that exist on different layers, and this fact is the principal reason why this type of rule is required. Whenever structures are to be manufactured using polygons on two different layers, there is a significant chance that there will be a misalignment between the desired and actual relative placement of the two polygons. Misalignment between polygons can result in both undesired open and short circuit connections, depending on the layers involved. Fundamentally, overlap rules reduce the impact of a small misalignment between layers in the manufacturing process by ensuring that the desired connectivity is maintained.

Let's consider an example where there is a contact between two interconnect layers. Physically, a contact polygon turns into a hole in the insulator between the two interconnect layers. The upper layer material must fill the hole and make contact with the underlying layer for the connection to be achieved.

The overlap rule states that the two layers in question must not only overlap each other; one layer must *surround* the other by a certain value. This value is the value for the overlap rule. In the case of the contact, the upper and lower layers must completely overlap the contact and surround the contact hole by the overlap rule value. If one of the layers does not sufficiently overlap and surround the contact hole, then the connection will not be reliable under all manufacturing conditions.

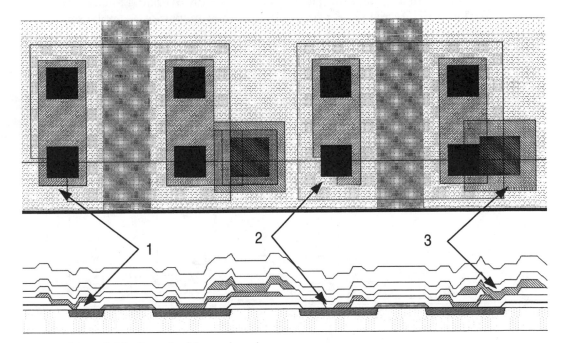

Figure 3.18   Example of the overlap rule.

What does the overlap rule achieve? In the case where the physical polygons are not aligned perfectly, there still will be enough upper material to fill the hole. If the upper or lower layers do not completely overlap the contact hole, the area that is available for the electrical connection is reduced. This results in a poor or weak connection (unreliable!).

In Figure 3.18, in examples 1 and 2, observe the result of poor contacts between active and metal1. If the active is not completely overlapping the contact polygon, the contact base is not wide enough. If the metal is not completely overlapping the contact polygon, then the contact hole is not completely filled and the contact will again result in a smaller connection surface area. In the third case we have an overlap problem between metal1 and metal2. The via has no metal1 overlap, so the connection, if any, is minimal.

The example in the figure demonstrates a case where an open circuit has a greater likelihood of happening. Let's consider another example where a short circuit is created when an overlap rule is not obeyed.

In this case let's consider different transistor layouts where gate polysilicon is combined with an active layer. To ensure that the transistor size is accurate and that a short circuit is avoided between the source and drain nodes, there are special rules related to transistors.

In general there are two overlap rules: active overlapping the gate and the gate overlapping the active areas. Figure 3.19 shows four different cases.

Node Out4 is an example of adequate overlap of the gate layer by the active polygon. Node Out4 is well defined. Contrast this example to node Out3. It is likely that the thin area of node Out3 will not be created.

Out1 and Out2 are examples of the gate layer overlapping the active layer. You can see that because the gate layer does not fully overlap the active area, nodes Out2 and VDD are shorted to each other, as they are part of one polygon of active.

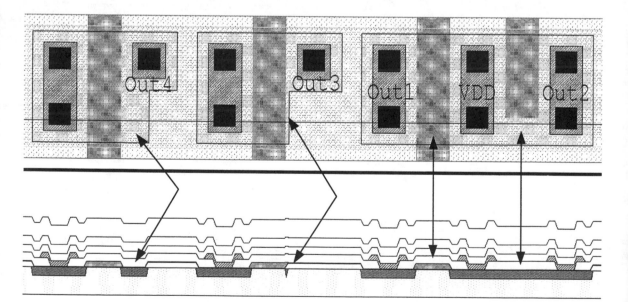

Figure 3.19   More examples of the overlap rule.

3.5   VERTICAL CONNECTION DIAGRAM

In the majority of ASIC processes there are very straightforward rules for con-nectivity. An ASIC process has one poly for the gate and three to six metals. That means that we have to work with one contact type and two to five types of vias. The connectivity is easy to understand because to connect from active to metal6 we need all the possible contacts and vias.

In more complicated processes, it is advisable to generate a vertical diagram of layer connectivity so the layout designer can fully understand the connectivity scheme. Figure 3.20 shows a DRAM process that is made with four poly layers and three metal layers. In this type of process there are restrictions on layer uses and connection rules.

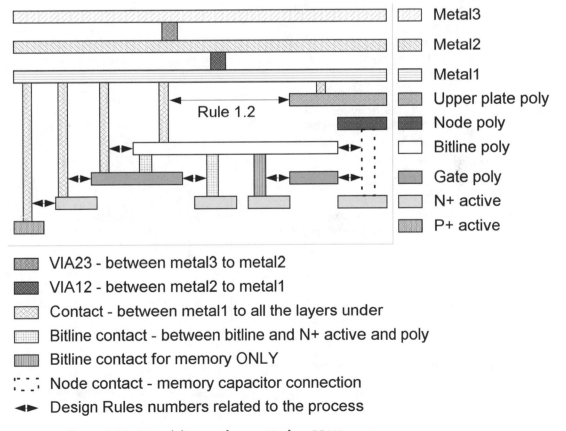

Figure 3.20   Vertical diagram of connectivity for a DRAM process.

3.6   A GENERAL PROCEDURE TO FOLLOW

Figure 3.21 shows a general layout design flow that is applicable for all design types.

This procedure is straightforward and self-explanatory and could be applied to almost any engineering task. Of course we will concentrate on how it applies to layout design.

Step 1 will be covered in detail, as it is crucial for getting started on the right track. It is in this step that we collect and review our knowledge of layout design and apply it to the specific circuit design under consideration. The aim is to produce a strategy for attacking the design by documenting the general areas where all components and signals will go.

Step 2 is simply implementing the design: executing and possibly revising the floorplan based on the realities of implementation. One way to think of the design process is to "plan top down," then "implement bottom up." By this we mean that we first floorplan general areas and approaches with a top-level view. With this plan in place we implement the design by starting with the lowest level components first and fill the areas defined in the plan. The lower level design tasks are easier because the constraints imposed on them were defined in the top-level floorplan. As the general areas are completed, we adjust our plan for future work. With a sound floorplan, adjustments are minor and the completion of the design is easy.

Computer-based checks (which will be covered in Section 3.10) form the bulk of step 3. These checks should be done in a certain order as outlined in Section

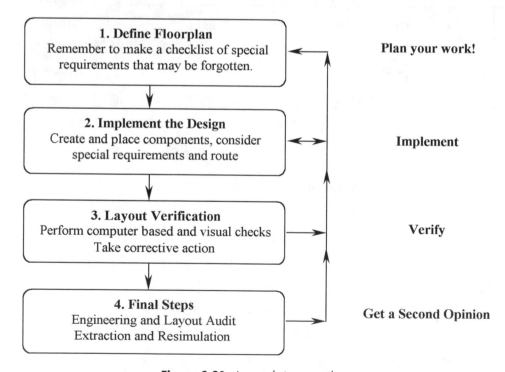

Figure 3.21   Layout design procedure.

3.10. On top of the computer-based checks a visual inspection is recommended, as the automatic computer checks are only as good as the rules that are coded into them. Make a plot of your design and look at it. Also, there are many aspects of most designs that cannot be checked by computer. An example of this is the degree of symmetry of a balanced layout. These visual checks should be part of the audit checklist as a reminder.

Step 4 is a final sanity and cross-check to confirm that all requirements have been met and none missed, along with a final extraction step.

One comment is necessary about the procedure and flowchart presented in Figure 3.21 and all procedures presented in this chapter. They all show that after each step in which the design is modified and evolved, it is necessary to go back to a previous step to readdress a requirement that may no longer be valid. Note that there are no shortcuts in the flow. All the arrows are going up *only* on the right side, and therefore all previous steps should be revisited after every modification.

3.7   PREPARING TO START

The most important stage in any kind of layout design is the *planning* stage. Quality in layout means that the end results, the final layout, meets the customer's (i.e., the design engineer's) requirements. To achieve quality results, a layout designer has to prepare a list of input requirements, taking into consideration all the specifications and a list of output requirements in order to ensure that output layout requirements are met.

3.7.1 Developing a Layout Floorplan

Now that we understand how to build single transistors and the concepts behind design rules and manufacturing process, we can start to plan our layout with some sound fundamental knowledge.

We need to remember the concepts presented in Chapter 2. The schematic or netlist that we have been given to lay out has undocumented or implied electrical and performance characteristics that need to be implemented for an optimal design. Ideally, a list of documented requirements is supplied with the circuit design. Ask for it!

We have one suggestion that can be of tremendous benefit at this point in the design process but is not indicated in the procedure. Depending on your familiarity with the type of layout you are about to start, it is always a great idea to do a bit of research to familiarize yourself with the circuit involved. It is extremely rare that you are about to attempt something that has never been done before!

Ask to see previous designs of the same type or in the same process. Ask who the expert is on this type of circuit. Review the concepts in this book if it has been a while since you last did a layout of this type. An appropriate amount of time to look for information to reuse and help you in your work usually pays off. Get a second opinion on your work (by reusing someone else's) before you even start!

Of course, the flip side of this approach is that it also pays to make your work and knowledge available to everyone else as well.

To ensure that nothing is missed, a prelayout checklist from the notes from your research is a great way to plan out a strategy for laying out a design. Figure 3.22 shows an example of a general procedure for creating a layout plan based on the circuit design requirements.

The first step, 1.1, is related to the planning of the layout of the power supplies and/or global signals. The power supply connectivity is typically called the power grid. Power supply resistance from the interface to all parts of the design must be considered. In this case special attention must be applied to the width of the supply lines and the grid or mesh of power lines through the design. Again the interface to other designs is important, especially in the case of a cell design where it may be desired to array it or have seamless abutment requirements to other cells. Let's not forget that tub and substrate contacts are typically connected from the power supplies, so a strategy to lay out these contacts must also be considered.

Step 1.2 is to list all of the input and output signals. Each signal is assigned a position on the interface of the design to the neighboring designs. The interface is defined as the boundary of the design. In some cases certain signals will have a specific or nondefault signal width assigned to them. Special considerations for signals may include clock signals, signal buses with multiple bits that need to be matched between them, critical path signals, and shielded signals.

In Step 1.3 we have to deal with special design requirements such as layout symmetry, specific requirements for latch-up protection, or noise immunity. More examples of special design requirements might be that the design must be pitch matched (i.e., limited in size in one direction), must have a very specific critical path signal, or might be a nonstandard part of the design.

Step 1.4 is very important to help finalize the size of the design and estimate the feasibility of meeting all of the design requirements within the area and schedule constraints. Using any previous knowledge about older designs of the same complexity and the process design rules, a layout designer can approximate the size of each component and the complete design. The number of different components to be implemented can be identified and the overall hierarchy or partitioning of the design can be completed. Areas for internal routing and signal connections should be allocated. The routing layer in each interconnect area should be identified. Extra signals or space should be reserved since we have only an educated estimate as to the size of the final design.

At this point we should have a preliminary floorplan and implementation strategy. The floorplan should comprise a definition of the interface or boundary with all of the signal ports assigned to their proper locations. Signals with special requirements are identified, and the area impact of these special signals is included in the total area estimate. If it is a hierarchical design, subcomponents are also known with their respective interfaces defined. Spare space and spare signal lines are included in this plan.

Step 1.5 is a sanity and cross-check to confirm that all requirements have been met and none missed. There are requirements related specifically to layout guidelines and styles for the process, but also circuit design requirements as well. The floorplan is a communication tool between the layout and circuit designer, as the circuit designer most likely had defined some specific requirements for his or her design and had assumed some kind of layout floorplan in modeling the design

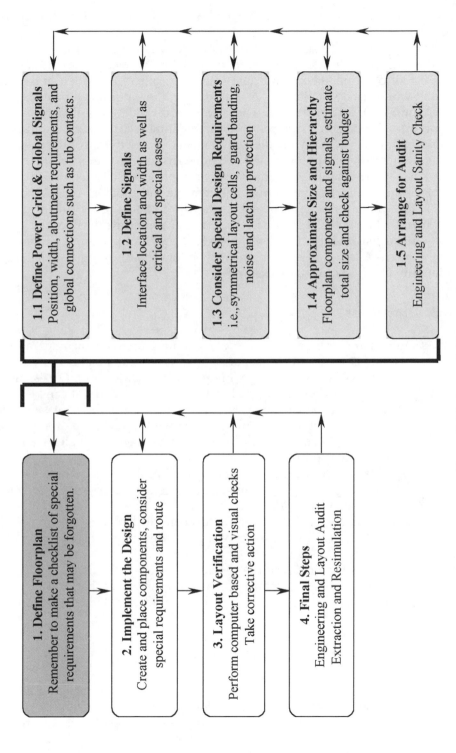

Figure 3.22 Layout planning procedure.

for its environment. It is very important to involve the circuit designer in the audit process.

The layout audit also relates to the next level of integration for the piece that you designed. The "brick" that you have now floorplanned has to interface perfectly to all of its neighbors and their interfaces; otherwise, when put together it won't work exactly as planned. As an example, the top-level chip design has to fit within the context of the chip package, so a review of any floorplan is very important. The person responsible for integrating your design should audit this floorplan as well to review the requirements related to size of the design, the layout architecture or approach, and your designs interface, among other things.

It is not uncommon for audits of really complicated floorplans to require two people: one to check the engineering requirements and another to check the layout needs. The auditor(s) should be a person who is not directly involved in the design, but who has the expertise to evaluate and appreciate the floorplan quality, and to make constructive comments. To help the auditor perform a proper audit, ideally a checklist is used for each type of layout design. Refer to the addendum checklists for some examples.

3.7.2 Stick Diagrams

Stick diagrams are a simple way of floor planning a circuit in preparation for layout. In many cases it is very useful for circuit design engineers to sketch for themselves a simple layout drawing without respecting any design rules, in order to imagine more realistically how the layout can be done and if it can be done at all. Circuit design involves many assumptions about how the final layout may look, so an easy and fast stick diagram can enhance the chance of a successful design. Stick diagrams can be done with various levels of detail that we will not go into here. An example is shown in Figure 3.23.

Step 1 shows a preliminary placement of all of the devices. In this case the VDD and VSS power line architecture has been predefined, as has the orientation and location of PMOS and NMOS transistors.

Step 2 shows the procedure for identifying which actives are connected to the same potential, and also the effect of flipping devices in order to take advantage of "sharing" these nodes.

Step 3 shows the final result of the active sharing. This is, in fact, the final layout with the interconnect optimized. The cell is narrower than in the previous steps. We can use this last version of the stick diagram as a floorplan for the audit mentioned in Section 3.7.1.

3.7.3 Hierarchical Design

As mentioned in Section 3.7.1, the circuit design analysis and resulting floorplan may indicate that it makes sense to have a hierarchical design. A *hierarchical design* is one that has a reference or uses another component as part of its construction. These subcomponents in turn may reference other components. This is similar to the concept of a subroutine in a computer program.

Building a design using subcomponents makes a lot of sense for the following reasons:

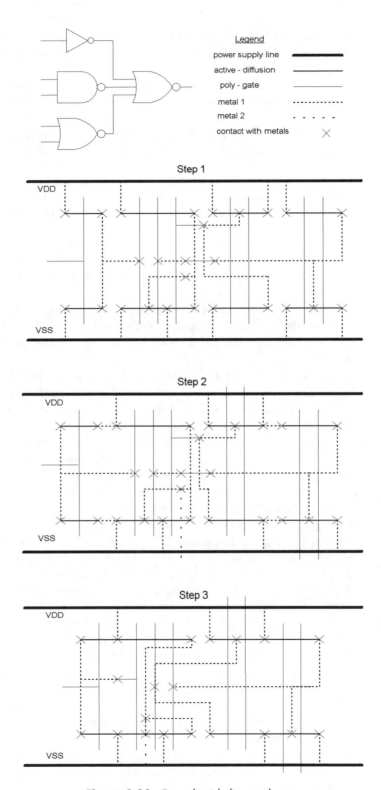

Figure 3.23 Example stick diagram layout.

- *Computer resource management:* Data that is already occupying disk and memory space is easily referenced as opposed to making a separate copy of the data.
- *Component reuse:* Designers can reuse components that are already fully completed—they have been designed, verified, and audited, preferably by experts in those areas.
- *Concurrent engineering:* Partitioning a design into different subcomponents allows many different tasks to be completed in parallel.

In terms of layout design we refer to these reusable subcomponents as leaf cells. The term *leaf cells* comes from the fact that a hierarchical design resembles a tree with a trunk, more primary branches, many more smaller branches, and finally, many, many, many leaves.

A few comments about leaf cells:

- They are repeatable layout designs that can be reused in many different regions of the chip.
- They can be made out of one polygon—i.e., contact cells; can be made out of three polygons—i.e., contact cells made with the surrounding layers (metal1, metal2, via12); or can be a complete circuit—i.e., inverter, NAND, flip-flop.
- They may have different versions of the layout for one version of the schematic because an inverter of equivalent size in the I/O region will have a different cell environment or interface than one in the memory region of the chip.
- Any group of polygons using a standard interface makes sense to be made as a leaf cell. For example, a library of logic gates all generally have a standard power supply layout interface, and so it makes sense for each of the gates to be a cell. If a design of multiple logic gates is to be implemented, it is not recommended that the design be done at the transistor level. Making logic gate cells first is preferred.
- If global changes are required to a design, it is much easier if cells are used. Imagine updating a design with 100 inverters. Consider the case of one design with an inverter cell instantiated 100 times versus another "flat" design with 200 transistors connected as 100 inverters. A change is required to all 100 inverters. In the case of the inverter cell design, the inverter cell is updated. Depending on the change required in the 200-transistor case it is likely that it would be more efficient to start the design over from scratch using cells than to try to update all 200 transistors.
- Conversely, using our 100-inverter example again, we must be very careful if only 1 of the 100 cells requires updating. We cannot change the inverter cell without affecting the other 99. In this case a second inverter cell is required that reflects the required updates and we replace the one outdated inverter with the new one.
- Every cell in a design needs a unique identifier even if it is a second instantiation of an already designed cell. In our example of the design with 100 instantiations of a cell called INV, we need to identify each one uniquely (i.e.,

INV001, INV002) and the identification of the cell should match the name that is used in the circuit design if there is an electrical correspondence. This *instance* name is needed to differentiate each of the physically identical INV cells. This is very useful in our example of changing 1 out of the 100 inverters, as we can use the instance name to identify the outdated inverter.

- Cells can be flipped and rotated much more easily than groups of polygons.
- If a symmetrical layout is required, one cell representing one-half of the design is all that is needed, and this techniques guarantees that the resulting layout is perfectly symmetrical!
- Cells can be scaled, although this is risky because of issues with polygon coordinates becoming off grid.
- Computer screen redraw resources is minimized using cells as all polygons within a cell need not be shown. It is often necessary to show only the key interface polygons and leave the details of the cell hidden.
- Cells can be "arrayed" to save more computer resources. An array can be thought of as the definition of a matrix of cells. For example let's consider the case of implementing a 10×10 matrix of memory cells. We are given a single memory cell as the leaf cell. One option is to instantiate the memory cell 100 times (this results in 100 X,Y coordinates, or 200 numbers that not only must be stored—we have to generate them as well!). A better option is to define an array that requires an X,Y origin, an X,Y offset, and the number of rows and columns (six numbers!).
- The use of cell arrays also reduces computer screen redraw time. For example, certain software packages have options to display only the border cells of an array. Another option would be to display only the corner cells.
- Hierarchical layout verification tools can take advantage of the repeated use of a smaller number of cells versus many individual ones. In our example of a design with 100 instantiations of a single inverter cell, conceptually a layout verification tool needs only to verify the inverter cell once and then check how each of these cells interfaces to each other. This approach is much faster than verifying the case of 200 transistors connected as inverters.

Figure 3.24 shows examples of cells instantiated in a design with instance names and different orientations. The stylized letter F attached to the cell origin shows the layout designer the orientation of the cell.

From Figure 3.24, please note the following:

- Base cell name = AGBC in all cases
- PDC is the name of the top hierarchical cell/block
- Instance names are different every time a cell AGBC is instantiated in the design, e.g., comp001, comp002, comp003
- The first example shows all the cells are oriented at 0 degrees relative to the origin of the PDC block
- The second example shows the instance comp002 is flipped versus the other cells—observe the two arrows that define the mirror axis
- The third example shows an array implementation

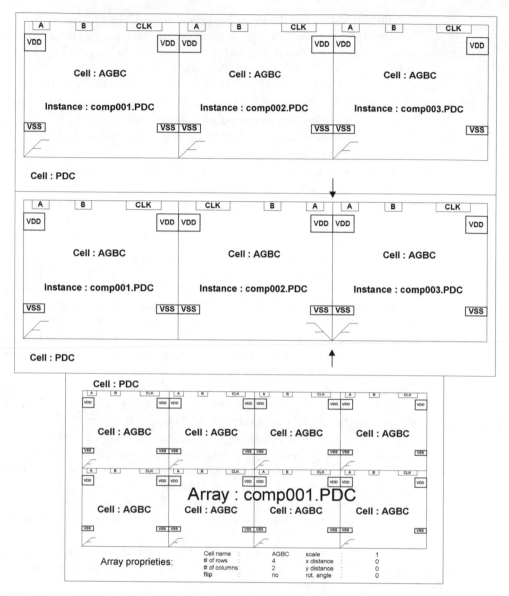

Figure 3.24 Examples of hierarchical designs.

3.8 GENERAL GUIDELINES

Now that we have considered a layout floorplan, it is time to implement the design. There are general guidelines that should be followed both in planning and implementing the design, and the fundamental ones are listed in this section.

3.8.1 Guidelines for the Layout of Power Lines

Power supply lines have to be determined before starting the layout of any cell. Specific guidelines for power lines include the following:

- Determine line width based on the following:

  Does the line provide power only for internal use, or is there a requirement for the line to feed other cells and be a part of the chip's power grid? This is critical information that can be determined from the layout floor plan.

  Use resistivity information of the different layers to determine the appropriate width of the lines.

- Use lowest level of metal for power for transistor level cells. It is important to consider that using higher layers of metal for power requires vias and local interconnect polygons to connect transistors to supplies. This will take space and limit the porosity of the cell. Generally use the lowest level of metal for power that the process and design allows.

- Avoid notching power lines. Power lines carry large amounts of current; therefore, it is important to make sure that they are routed with a consistent width and never notched. Any notches in a line create potential fuses that may break the power line under high current conditions.

- Avoid over the cell power routing. Unless the power is routed using automated tools, it is not advisable to run power supply lines over the cell. Keep the power lines inside the cell to ensure that the cell is correct by construction. Power routing outside the cell is ideally limited to connections between cells.

3.8.2 Guidelines for the Layout of Signals

The following is a list of guidelines specifically for routed signals. More details on each of these concepts are presented in later chapters.

- Choose routing layers based on process parameters and circuit requirements. For each process a standardized list of routing layers should be determined based on layer resistance and capacitance. Layers such as N-well, active, and high-resistance poly gate are not used for routing. Priorities between routing layers can also be standardized using the same criteria.

- Minimize the width of input signals. Minimizing the routing of the signals within the design is also important. This reduces the input capacitance for the signal. This is important for signals that are part of cells that are used many times. An example of this would be a clock signal within a cell.

- Choose routing width carefully. The choice of the width of a routing signal should be made judiciously. It is tempting to simply use the minimum design rule line width as the routing width. This is not practical in all cases because connections must be made to every line.

  These connection points require a via or contact, and as you can see in Figure 3.25, the space required for a via or contact is generally wider than the minimum width rule for the routing layer.

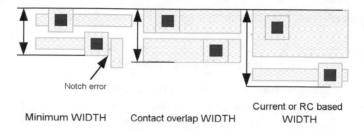

Minimum WIDTH   Contact overlap WIDTH   Current or RC based WIDTH

Figure 3.25 Routing line width considerations.

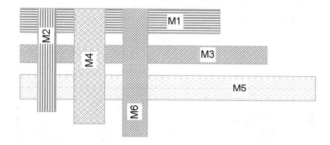

Figure 3.26 Routing direction.

It is perhaps more efficient to use a width of line that accommodates both a via and contact. This line would be wider but is much easier to manage and maintain as it avoids jogging other signals around the contact point.

- Maintain a consistent routing direction within a cell or block (Figure 3.26). Then, when a layout designer wants to change signal direction, it will take only one type of via and it will not interfere with the porosity of the cell. In general it makes sense to keep a consistent metal direction for each layer and alternate the direction from layer to layer. For example, if metal1, metal3, and metal5 are routed horizontally, then metal2, metal4, and metal6 should be routed vertically.

- Label all important signals. This is very important for the layout verification process, especially LVS. Error diagnosis, short isolation, and LVS run time are all easier when nodes are labeled.

- Determine the minimum number of contacts for every connection. Don't assume that a single contact or via for every connection is enough. Some memories, for example, are using double contacts everywhere possible to increase reliability.

3.8.3 Guidelines for the Layout of Transistors

The following is a list of guidelines specifically for a cell-level design environment.

- Use a predefined template for PMOS and NMOS transistor placement. The architecture of a cell should be defined beforehand, and this template should encapsulate the basic floorplan of a group of cells. Figure 3.27 shows an example of a cell template.

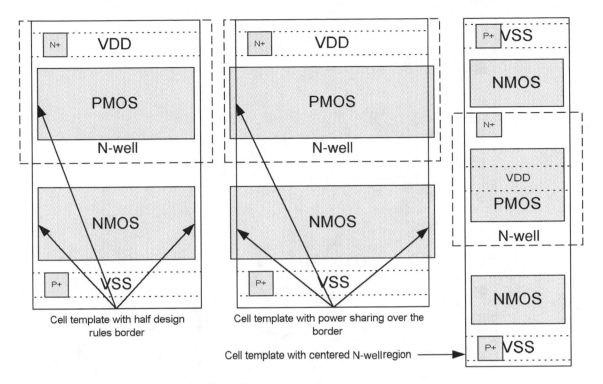

Figure 3.27 Example cell template.

- Use transistor fingering for large and critical transistors. A cell template similar to Figure 3.27 defines the maximum width of a transistor by the cell height. How do we lay out a transistor that exceeds this height? The solution is to "finger" the transistor into multiple transistors that are connected in parallel.

  Figure 3.28 shows three equivalent layout designs of a transistor that is 100 µm wide.

  There is an advantage with an even number of fingers: the active capacitance is less, because the drain region is surrounded with gate poly instead of field.

  Another reason to use fingering is to optimize the resistance of the gate poly along the width of the transistor. Since the gate poly is driven from one end and gate poly is resistive, there may be reason to have a guideline that states the maximum width of a single finger. Fingering is the only way to meet this guideline for large transistors.

  Fingering multiple transistors that are connected in series is trickier. Figure 3.29 shows an example of fingering a two-input NAND gate.

  Fingering the PMOS devices is straightforward; however, fingering the series NMOS devices is more difficult because the order of connectivity of the devices must be maintained.

- Share power supply nodes to save area. Sharing nodes whenever possible is a concept that is easy to understand. Power supply nodes are most easily

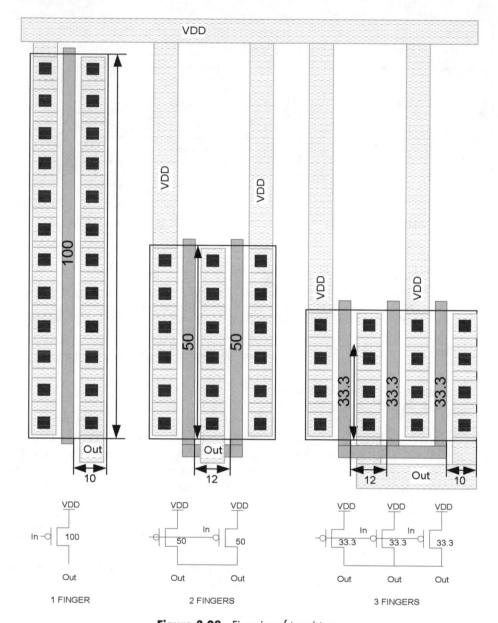

Figure 3.28 Fingering of transistors.

shared because they are very common and easy to connect. Very significant area savings can be achieved.

The main reason that area is saved is that both sides of a row of contacts are used and there is no need to space two active regions apart from each other. Figure 3.30 illustrates this technique.

Note that power nodes can be shared between transistors of different widths with a slight overhead of inserting a poly to active space at the end of the smaller transistor.

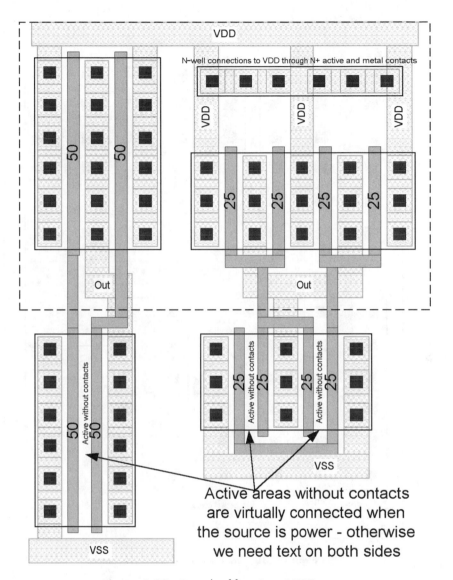

Figure 3.29   Example of fingering a NAND gate.

- Determine minimum number of contacts for source and drain connections. One simple rule might be to fit as many as you can using the minimum design rule between two contacts. This guideline is most reliable and maximizes the performance of the transistor. The downside of this approach is that the routability over the transistor is limited.

  If increased routability is required and accounted for in the circuit design, fewer than the maximum number of contacts may be optimal for the overall layout design. This approach must be considered carefully and accounted for in the circuit design process.

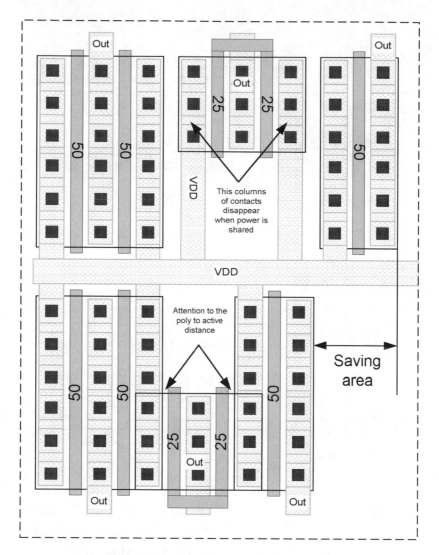

Figure 3.30 Transistor power sharing example.

In general, ASIC cell libraries use a minimum number of contacts within the cells, but for some high-frequency designs or analog parts, the transistors are fully contacted.

• Use 90-degree polygons or paths whenever possible. Most designs are done this way. The reason is that with orthogonal shapes the computer has to store a minimum amount of data and the layout process is much easier.

You should reserve 45-degree layout design for areas that have tight area and performance constraints. The reason is that 45-degree layout is more difficult to modify and maintain. The extra effort to use 45-degree layout design techniques is not worthwhile for the average layout cell.

Examples where 45-degree layout design is worthwhile are for memory cell and pitch limited layout, as well as datapath and large power supply

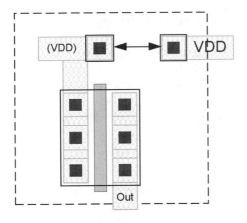

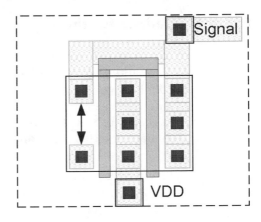

N-well Soft Connect   Transistor Drain Soft Connect

Figure 3.31 Example of soft connections.

lines. However, 45-degree contacts and vias cause many problems for many CAD programs, so they are to be avoided.

- Plan for and standardize tub and substrate connection locations. Plan and place connections of the N-well to the logical "1" power (VDD, VCC, VPP) and the P+ substrate to ground or VSS.

    The floorplan of the cell should include the general area where these contacts are to be placed. There are two basic methodologies: to place them between PMOS and NMOS transistors, or to place them on the outside of the transistor region. Between the transistors is better for latch-up protection (to be discussed later), but this complicates a cell layout.

- Avoid "soft connected" nodes. A "soft connected" node is one that has been connected through a nonrouting layer (Figure 3.31). Nonrouting layers are usually identified as such because they are highly resistive and result in poor circuit performance.

    Typically, active and N-well layers are not routing layers, but it is still possible to inadvertently use these layers to make electrical connections. The problem with this is that the DRC and LVS will pass, but the circuit performance will be poor. Only a very detailed layout extraction and simulation will find this type of "soft" error. Typically, this type of work is not practical, so a correct-by-construction approach is taken to avoid this effect.

    Figure 3.31 shows two examples of soft connections. The N-well example shows how a transistor is electrically connected to VDD, but the signal path flows through N-well as part of the connection. The second example shows how the transistor performance may be compromised by a connection to the drain that is not completed in metal. In this case the single contact does not help in any way, and an equivalent layout would be one without it.

    Special checks built into the layout verification process can help to identify these problems; however, they are difficult to debug, and it is best to simply avoid making the error in the first place.

3.8.4 Guidelines for the Layout of a Hierarchical Design

- Develop and use a floorplan plan. This cannot be emphasized enough, and it should be done at all levels of layout design.
- Define the hierarchy of a design in the planning stage. There are no hard-and-fast rules for defining the hierarchy of a design, but common sense is hard to beat. Common guidelines for determining different levels of hierarchy include the following:

  Circuits that are to be instantiated many times need to be cells

  Divide designs into functional or area-specific blocks

  Divide designs into blocks that allow multiple designers in parallel to work on them

  If symmetrical layout is desired, use a single half cell and mirror it to complete the design

  The use of hierarchy is discussed in many areas throughout the book.

- Develop and obey standards for layout near the boundary of a cell. The floorplan and the type of cell that is being implemented should define how the cell should interface to its neighbors. The interface requirements of any design should be known and understood in the planning stage.

  Here are some techniques and guidelines to properly design the interface for a cell:

  Use template cells to define global characteristics—cell dimensions and the placement of power supplies and wells are good candidates to define the interface of the cell. Consistent use of templates ensures that all cells conform to a standard and will integrate together smoothly.

  Assume that a boundary interface is fixed—if any polygon is required to cross the boundary of the cell, then the floorplan at a higher level in the design needs to be consulted before it is allowed. This avoids overlapping polygons with those that are unseen.

  Half design rule approach—if the cell is to abut to itself or other cells with similar boundary conditions, then a correct-by-construction approach would be to ensure that all internal polygons are spaced away from the boundary by a value that is half the specific design rule. In this way, when the cell abuts to another cell, spacing rules will not be violated.

  Verify the cell with its neighbors—this technique guarantees that the cell is correct in all cases.

3.8.5 Quality Metrics

Once the essential requirements have been met and/or we become experienced in doing layout, the following list outlines more subtle topics and metrics that will be covered later in the book. We only mention them here to introduce the topics so that we can anticipate and potentially plan ahead for these advanced requirements:

- Area
- Performance
- Porosity

- Manufacturability (i.e., is it all minimum design rules? If not, then it could be considered more manufacturable)
- Maintainability (i.e., will the layout be easy to change or optimize if the process changes?)
- Reliability over the long term (i.e., electromigration)
- Interface compatibility (i.e., does it abut and fit well in all instantiations?)
- Shrinkability (i.e., does the layout lend it self to future shrinks of the process?)
- Reuseability (i.e., does the layout lend itself to migration or retargeting to different processes?)
- Compatibility to layout flow (i.e., is the layout friendly for all downstream tools and methodologies such as P&R?)

3.9 IMPLEMENTING THE DESIGN

Let us now put all of the knowledge we have learned so far to work. At this point we have all of the fundamental knowledge to complete a basic layout design.

First, let's review the key concepts presented so far (Table 3.1). If you have mastered these fundamental concepts, you should be able to tackle almost any layout task.

We have learned that before we start any layout design, we must make a plan. Before we go ahead and execute this plan, it is always important to anticipate what the next steps are and keep these in mind as we implement our design. In this way we minimize or eliminate the amount of rework caused by our own ignorance.

Component placement (Step 2.1 from Figure 3.32) based on the floorplan is another area where we will always achieve a good return on the effort invested in doing a proper placement of components. The ability for the design to be completely routed is usually limited by the placement of components. These components can be other transistors, tub contacts, power supply lines, or interface

TABLE 3.1 Concept Review

Chapter	Concept Summary	Comment
1	What is layout design?	Also covers how layout design fits in the IC design flow
2	How do I read a schematic?	A schematic has more information than meets the eye!
3.2	Layer definition	An introduction to CMOS processes
3.3	CMOS transistor layout	A basic introduction
3.4	Design rules	These define the limits of what you are able to do
3.5	Layer connectivity	This defines what can be connected to what
3.6	A procedure to follow	General instructions
3.7	Developing a plan	Potentially the most important step for success!
3.8	General guidelines	Concepts to follow to do it right the first time

locations for signals, in addition to instantiated components. Any extra effort in the placement stage will be rewarded in the long run by having an easily routable design.

It is at this stage that signal lines or interfaces should be labeled and identified to avoid connection errors. This would include power, signal, feed-through, and other polygons.

Step 2.1 can be considered to be an initial placement to validate that the floorplan is feasible. In Step 2.2 we start to consider finer details, where we have to deal with special design requirements such as the critical path signals, substrate contacts, layout symmetry, specific requirements for latch-up protection, or noise immunity. Detailed placement will occur in this step as well as the routing of important or difficult signals. Power supplies and clock signals fall into this category. In addition, extra space for components and routing should be allocated in anticipation of new design requirements after the layout has been started or completed.

With a good floorplan and final placement of components, Step 2.3 becomes very easy. Without the former, completing interconnect routing while respecting special design requirements is usually difficult and time consuming. Remember that the floorplan should have considered the routing layers, the direction of routing, and the space for routing all signals, so by this time completing the routing should be fairly straightforward.

Let's now examine issues related to specific types of layout designs.

3.9.1 Cell Layout

The leaf or logic cell is in general a layout that is drawn from a transistor-based schematic; therefore, the majority of components to be used in this type of design would be polygons, transistors, contacts, and signal pins. "Polygon pushing" is the layout design style used here, as we are implementing circuitry at the lowest level of abstraction and we need detailed knowledge of the entire set of layers and layout design rules. Formally, this is known as a "full custom" design style.

The key concepts to be addressed at this level of layout include the following:

- Detailed knowledge of the entire set of layers and layout design rules.
- The size of the design, estimated from the number of transistors in the design and the layout design rules.
- Attention to transistor-level placement and interconnect to implement logic gates.
- Careful floorplanning and architecture definition to minimize area and maximize performance. These leaf cells are potentially used thousands of times, so the extra effort in achieving area savings for each cell is justly rewarded in the finished chip.
- Careful design of the power supply implementation. This also includes consideration of substrate and tub contacts. If this is done well, the power supply routing and bulk connection requirements of an entire block or chip can be met by building these requirements into the design of the leaf cells.
- Attention to the design of the interface to other cells. As mentioned previously, these leaf cells may be used many, many times, and area savings can

be achieved by minimizing the overhead required to place two leaf cells adjacent to each other. Ideally, leaf cells should be designed to abut directly to all possible cells that may be placed adjacent to them.

3.9.2 Block Layout

The difference between a cell and a block is open to interpretation, but in general a cell is referred to as a block when it incorporates circuitry of medium complexity and functionality and is mainly composed of instantiated cells. Blocks are larger and more complex than cells and are implemented by designers with more experience than those who can design cells. We consider this design style to be a "semi-custom" design style, as it combines cell-based design with full custom design. The layout flow described in Figure 3.32 still applies.

The important factors to keep in mind for block level design are as follows:

- The size of the design is estimated from the number of cells in the design and the number of external and internally routed signals.
- It is most common in blocks to have a significant amount of space allocated for spare components and signals as well as signals that may simply pass through the block. These signals are referred to as feed-through signals.
- Some blocks will have functionally different components such as a mix of digital and analog cells. In this case there may be special considerations for the different circuit requirements such as latch-up and noise immunity.

In general, when the layout designer has all the above information, it is time for placement of the previously made cells, transistors (if any), and interconnect routing.

In general, blocks are defined based on the circuit functionality and the layout style. Examples that are common might be row cells, standard cells, datapath or register files, I/O (input/output) ring, analog blocks, and memory blocks.

3.9.3 Chip Layout

Implementation of the full chip layout is conceptually identical to that of a large block in that the steps from layout planning through to auditing remain the same.

Full chip layout designs will incorporate blocks and cells of diverse types, so detailed knowledge of all layout design styles is required to maximize the chance of success.

Aspects that must be addressed at the chip level and generally nowhere else include the following:

- Design partitioning of very large and diverse blocks. Defining the many interfaces of a complex chip is a complicated task.
- Defining and planning the interface to the outside world. This involves knowledge about pad and I/O circuitry and the intricacies associated with them.
- Planning and implementing critical signals that are routed over the entire chip. This would include global power supplies and clock signals.

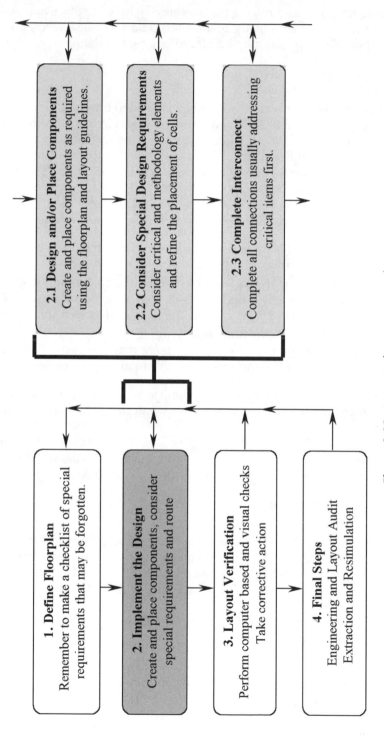

1. Define Floorplan
Remember to make a checklist of special requirements that may be forgotten.

2. Implement the Design
Create and place components, consider special requirements and route

3. Layout Verification
Perform computer based and visual checks
Take corrective action

4. Final Steps
Engineering and Layout Audit
Extraction and Resimulation

2.1 Design and/or Place Components
Create and place components as required using the floorplan and layout guidelines.

2.2 Consider Special Design Requirements
Consider critical and methodology elements and refine the placement of cells.

2.3 Complete Interconnect
Complete all connections usually addressing critical items first.

Figure 3.32 Layout implementation procedure.

- Floorplanning techniques and maintenance are of paramount importance here. At the full chip level the floorplan is critical as a communication tool as well as a layout implementation tool.

- Estimating the chip size is a significant task in itself. Estimates from previous designs and previous experience should play a role in completing this task. Compare the process parameters of the current project to previous ones. This is one area in which expertise in floorplanning tools can really help.

- In the role of a layout leader responsible for a full chip layout, there is also the requirement to define layout methodologies, task allocation, and scheduling for the entire team. Also, an understanding of the suite of layout design and layout verification tools is important in ensuring that the team performs efficiently.

Overall, it is the complexity of the task of the full chip layout that makes it one of the most challenging and interesting roles in layout design.

3.10 VERIFICATION

Now that the implementation of the layout is complete, we move to the verification step (Figure 3.33). This is not a small task, and it is a very important one. There are many failure mechanisms in IC design, and fixing errors is very expensive. Unlike fixing a car, where access to components for replacement and modification is relatively quick and easy, fixing design errors can take months. We should take the approach that we only have *one* chance to get our design right, because a revision to a design is a very lengthy and costly process (somewhat like trying to fix a satellite after it is in orbit—possible, these days, but extremely costly).

In spite of all of the planning and checklists, a robust verification plan is required for best results. Each step in the plan checks different aspects of the design.

3.10.1 Design Rules Check (DRC)

The design rule verification step checks that all polygons and layers from the layout database meet all of the manufacturing process rules. As described in Section 3.4, these design rules define the limits of a manufacturable design. Width and space rules fall into this category.

Meeting the manufacturing requirements is the absolute minimum rule set that must be checked and corrected. Because this is the first level of verification once the layout is implemented, typically many methodology, connectivity, and guideline rules are checked as well. We refer to these as a set of supplementary rules. An example would be an illegal use of layers (ESD layer in the logic area) or illegal devices or connections.

Tip: A truly complete DRC verification approach would be to verify not only the design that you, as a designer, have implemented, but also your design placed within the context in which it is going to be used.

If the specific components that will interface or be adjacent to your design are available, perform a DRC check with this interface cell included. If your cell is a general-purpose design, then a more intricate and exhaustive check should be performed, perhaps including all possible interface cells as well as different

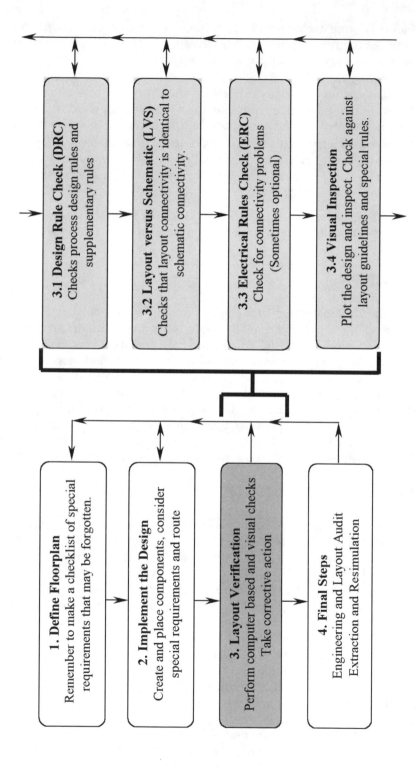

Figure 3.33 Layout verification procedure.

orientations and combinations that may occur. These approaches really eliminate the possibility of errors as your design is integrated into the overall chip.

3.10.2 Layout versus Schematic (LVS)

LVS verification is checking that the design is connected correctly. The schematic is the reference circuit and the layout is checked against it. In principle, the following is verified:

- Electrical connectivity of all signals, including input, output, and power signals to their corresponding devices
- Device sizes: transistor width and length, resistor sizes, capacitor sizes
- Identification of extra components and signals that have not been included in the schematic; floating nodes would be an example of this

The last item overlaps into the items checked in the electrical rules check, which is described next.

3.10.3 Electrical Rules Check (ERC)

As noted in Figure 3.33 the ERC is sometimes an optional or seldom used as an independent verification step. Many of the issues are caught by the LVS check, and thus the ERC has become redundant.

Electrical rules checked in this step are usually limited to errors in connectivity or device connection. Examples include the following:

- Unconnected, partly connected, or extra devices
- Disabled transistors
- Floating nodes
- Short circuits
- Special checks not checked elsewhere (i.e., antenna rules)

As a subset of the LVS check, an ERC generally executes more quickly and therefore is useful to accelerate debugging problems such as a VDD-to-VSS short circuit.

3.11 FINAL STEPS

Despite tremendous advances over the past 10 years in tools for IC design, and despite the best intentions of the top designers, there is always room for improvement and opportunity for something to be overlooked. The complexity and variety of process design rules over time also creates new challenges to old problems. For these reasons, a final sanity and cross-check in the form of an audit is time well spent (Figure 3.34).

Compared to many, many years ago, audit checklists today may be smaller, but the burden has simply shifted to the personnel responsible for the setup and maintenance of the layout CAD tools and layout flow. CAD verification tools

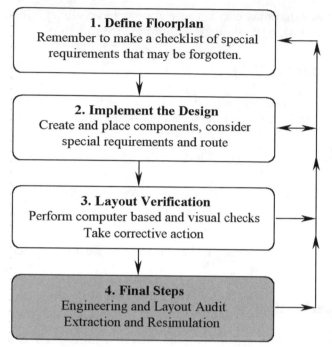

Figure 3.34 Final steps.

check many issues, but it still takes a knowledgeable person to ensure that the tools are doing their job.

The secret of proper audit results is to have an auditor who understands the concepts behind the issues and the extent of any problem, and who can also propose solutions. It is a good idea to have a third-party auditor: a person who has not been directly involved in the project or design being audited. This eliminates any predisposed bias or assumptions.

The procedure is straightforward. The auditor should review the requirements and documentation for the design and, using checklists specific to the type and complexity of the design involved, verify the final design against these requirements. The checklist generated in the floorplanning step is a prime example of documentation for the design.

Issues that are raised are documented. At this point it is a good idea to involve the circuit designer in the audit process as well as the person responsible for integrating the design into the next higher level. This way a solution to any problem identified can be dealt with efficiently and with all the relevant information at hand.

All issues are signed off by the auditor who identified the problems, once the solutions have been implemented.

The very last step in the process is to generate an extracted layout.

Extraction is a hand-off step back to the circuit designer. A version of the final layout design compatible for simulation is given back to the circuit designer for final resimulation. Tweaks to the design may occur after simulation.

Extraction is the process of automatically generating from the layout a netlist compatible for simulation that includes information that corresponds with all the device connectivity, device sizes, and routing capacitance and resistance.

The extracted netlist is a good communication mechanism between the layout and circuit designer. It also indicates that the layout design is complete pending the final simulation result.

3.11.1   Verifications

In terms of verifications of the final files, here are some considerations.

Because the mask shop requires GDSII file type, the final verification will be done on the *same* file that goes to manufacturing. If the layers shipped to the mask shop are different from the ones used for design, the final verification has to be done on the GDSII required by the mask. The database has to be translated from the design layers that were used and verified online into the mask shop layers and verified as the final "golden" verification. However, there are a few structures that we add for the processing needs that won't pass DRC or LVS verifications without errors. These structure-generated errors must be checked very carefully, and not ignored, because they can touch real sensible logic circuitry, and then we have a problem.

Most important factor is that the final verification be done using "frozen" GDSII generated from a "frozen" database, which means that nobody can touch the original online data. This way you can ensure unique data.

3.11.2   Audits

At this level audits have to be performed by an experienced person who has already passed through one or more tape-outs. As a baseline check, the layout designer can use the checklist provided earlier, adding any company- and process-related questions. The other important issue is to audit the newly placed keys that may not conform to the DRC command file. (Refer to the checklist.)

A very important task, that is sometimes forgotten, is to re-audit the design if the command files for DRC and LVS include verifications for layers that are not designed in layout but generated using CAD software before mask making. One example is the N+ layer that surrounds the N transistors placed on the P+ substrate. In general, it is not drawn by layout designers and is generated only before mask manufacturing. In our experience, we have found that layers like this can generate problems if they are not checked before mask preparation.

3.11.3   Tape-out Procedures

When people talk about tape out procedures they are referring to the steps detailed above plus specific documentation sign offs and release procedures. For each chip released to the mask shop, there are internal company procedures that have to be followed. For example, the division management, together with the project leaders and the circuit and layout designers, should review all audit reports, additional company standard sign off procedures, and sign and release the tape and the accompanying documents and data.

It is a good practice to take care that all the macros, documents, verification results, audit reports, and command files related to the chip that has been taped out are filed for easy access in future releases. If there are any problems with the released project, it will be easy to access all the related setups and documents.

CHAPTER FOUR

Layout Design Flows

In Chapter 1 we defined what layout design is and explained where layout design fits in the overall process of IC design. The term "flow" was used many times throughout the text.

In this chapter, we will try to define what a flow is, why there are many different flows, how all these flows were developed, and what are the most common types of flows today. Examples of the most popular design flows today will be explored. We will close this chapter by summarizing the three basic design flows and the specific types of tools used to address their needs.

4.1 WHAT IS A FLOW?

We define a flow as follows:

> *An effective methodology of capturing and verifying a useable representation of an idea such that the final result exhibits the appropriate characteristics for its intended use.*

A *methodology* is "how" to do something. A methodology defines a series of steps to follow, but also includes knowledge about what issues to be aware of, tricks of the trade, and other goodies as we pass through these steps.

An effective methodology is one that is appropriate for the task at hand. This is important to the validity of any design flow. The flow that is used must match the requirements of the style of design that is being attempted; otherwise, failure is certain.

When is a flow effective?

The absolute minimum requirement for any design flow is to be able to *capture a useable representation of an idea*. A useable representation of the idea is the key concept to be understood. An idea can be captured on paper, but it is not useful other than for communicating the idea. The data must be useful and be a functional representation of the idea. A computer database of schematics, block diagrams, netlists, and layout drawings is the useable representation of the idea that all IC design is based on.

Without this minimum level of functionality we do not really have a flow at all, so this does not really answer the question of the effectiveness of a flow. If we must choose between two flows that can capture and implement our idea, how do we determine the effectiveness of each?

The more effective flow is the one that can produce a design that *exhibits the appropriate characteristics for its intended use* in the shortest amount of time or for the least cost.

Characteristics include issues you would expect in a formal specification of the design (Table 4.1).

It is no wonder that designer expertise is still in high demand, as it should be apparent that CAD tools cannot address all of the issues listed in Table 4.1. In fact, there has been a proliferation of tools as new vendors and new tools try to specialize and solve some of these issues individually that were previously not addressed in existing flows.

Appropriate characteristics *for its intended use* implies that our design need only "meet the requirements" for each characteristic, and not necessarily exhibit the "smallest or fastest" characteristic that is humanly possible. If a flow produces a sports car, but a hatchback was required, then the flow is not effective even though all performance specifications were exceeded.

How is the design's appropriateness measured as the flow is executed? The characteristics are constantly *verified* and measured at each step in the flow.

In a competitive marketplace it is ultimately business issues such as revenue and market share that determine the success or failure of the product. This is where an effective flow can amplify differences between companies. Some companies will produce an equivalent design more quickly than others (thus gaining market share before anyone else), others will produce a design that exhibits superior characteristics (such as speed), and still others will produce a design for a lower cost (thus being more efficient).

In summary, a flow is an encapsulation of knowledge of how things should be done. A flow defines a sequence of steps and a set of design tools for a specific type of design style. Successful companies are able to evolve and optimize their design flows in step with the business requirements for delivering their products. In the area of layout design, this book will explain the fundamental principles that form the foundation for any design flow.

Historically, all the designs were done in a full-custom fashion, meaning that each piece of the project was hand designed, verified, and laid out. As soon as the first computers started to provide proper platforms and specialized software for

TABLE 4.1 List of Design Characteristics

Behavior	Timing	Power	Area	Interface
Process	Maintainability	Reliability	Porosity	Flow compatibility
Complexity	Noise immunity	Yield	Electromigration	Latch-up immunity

IC design, developers of new products started to adapt their design flow to the new tools. In the past, methodologies and flows were dependent on the type of the design (i.e., full-custom, ASIC, memories), but more and more, they have become a mix and match of different flows.

We will try to define the most important changes over time of the standard flows and our perception of the reasons flows had to evolve:

- *Time to market pressures:* An integrated flow, even a full-custom one, is far more efficient than no flow at all. A flow is an approach that promotes faster design and proper allocation of tool resources, and most importantly, it shortens the time to tape-out. For example, a chip of 10,000 gates was designed in 6 months as compared to an earlier chip which took 2 years to design.

- *Increasing chip complexity:* A full-custom flow and tools cannot enable a team of designers to produce a design of 100,000+ transistors fast enough, even with development of faster tools and computer hardware. Chip designers had to develop not only new chips, but also new methodologies to cope with the increased complexity of the designs.

  The solution was to implement the designs at a higher level of abstraction and trust the design tools to work with more abstract models and reduce the time spent for the design of each transistor. HDL-based design methodologies matured at this point and enabled logic circuitry to be designed at the RTL level instead of the transistor level.

- *Growth of design services and fabless semiconductor companies:* The next market reality was that many experienced designers decided to leave their big companies and go out on their own. Startup companies in general were trying to address markets that were too small or were not understood by the big companies.

  As fabless companies, these startups designed chips and had their ideas manufactured by semiconductor companies under a business model now known as an ASIC model. These new companies demanded and bought new tools and methodologies to support their business.

- *Increasing chip size and shrinking device geometries:* Everybody tried to improve profits and to fit more and more transistors on a chip, so the process technology has been constantly evolving. The transistor gate size moved from 5 to 0.5 µm in less than 10 years and from 0.5 to 0.18 µm in the following 5 years. Using such small dimensions for their devices, chips can support up to 10,000,000 transistors for a microprocessor and 256,000,000 for a DRAM.

  The small transistors, capacitors, and resistors mean that many analog and physical effects influence the performance of the designs. Logic designers now have to contend with effects that previously only analog or RF/microwave designers had to deal with. An emphasis on signal integrity and accurate layout extraction tools is a major concern within design flows today.

- *System on a chip designs:* The large chip sizes have enabled the ability to integrate many different applications on a chip for more efficient system design.

This is referred to as the concept of *system on a chip* (SOC). No single tool or platform can be used to design a chip with such complexity in a short time without having huge design groups that are specialized in all the various applications.

Instead of designing the many applications internally, a new market of selling and buying different functional blocks has evolved. These blocks are referred to as intellectual property (IP) blocks, and this business approach promotes design reuse and is intended to increase the productivity of the design teams.

In this case, a design team will develop specifications for a full chip, but will design only the blocks that are in their area of expertise. For all the other blocks they will import, at different levels, blocks already designed and proven functionally by other teams or companies.

These are a few examples of the important revolutions that changed the way people design VLSI chips. Design teams have been required to adapt to market and schedule pressures, knowledge limitations, and changing process capabilities; new flows and methodologies are key to ongoing success.

What determines the flow that is used, or how is a flow defined? The answer is that it is the type of chip that is to be designed and the different blocks on the chip. The flow may be different for different blocks on the chip, especially for SOC designs. Microprocessors require one design flow for timely completion, while memories required another. Common products that are recognizable to most and their respective flows will be discussed in the following sections.

4.2   MICROPROCESSOR DESIGN FLOW

The most famous microprocessors are the Intel Pentium series, Digital's Alpha, and the workstations' RISC chips. Otherwise, small microprocessors are widely used as control devices for household appliances, business machines, cars, toys, etc.

An MPU, or microprocessor unit, is an extremely complicated integrated circuit that accepts coded instructions, executes the instructions received, and delivers the requested result.

Microprocessors are the most complicated chips, so in general every step is very well defined and checked at high levels of abstraction. Only after the concepts are proved are they entered in a normal full-custom design flow. The following is a very simplified example of a development flow for a microprocessor (Figure 4.1).

1. *Chip functional specification:* The first step is to define from marketing information the kind of functionality that is required for the chip. Based on the blocks and possible applications, the project leader decides to pursue a known standard architecture or develop a new one.

   The chip must comply with standards for bus interfaces, timing specifications, and manufacturing packages; therefore, the project leader has to decide the order of importance and priorities for each of them. Generally,

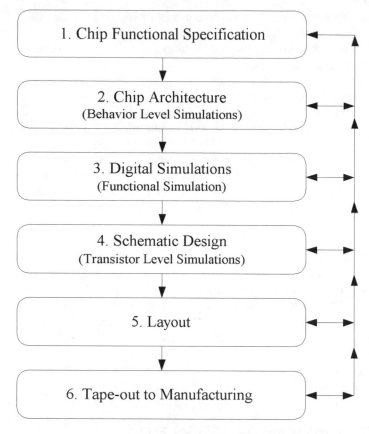

Figure 4.1   Full-custom flow for a microprocessor.

not every possible requirement can be met, because it would take too much time to address them all. After all, a chip that is perfect but does not come out in time for the market does not bring revenue to the company.

2. *Chip architecture:* The architecture of a chip is essentially the way the chip is broken down into functional blocks and how the blocks are connected together. In defining the architecture it is the choice and definition of different blocks that is optimized to meet the desired performance goals.

   In a layout sense, the architecture denotes where and how big each block has to be, where it makes sense to place the blocks, what the logical relationship is between them, and what the important signals are, among many others. At this level the chip looks like a chip with numerous empty blocks inside surrounded by a frame of pads.

3. *Digital simulations:* Digital or functional simulations are performed as part of validating a behavioral model of the intended design. The simulation verifies that the chip architecture is feasible and will perform the desired operation.

4. *Schematic design:* At this point the different blocks are refined as schematics that represent the design to the transistor level. At this level the designer has

to deal with real-world phenomena, such as dissipated heat, power consumption, resistance, and capacitance of devices and lines. Functionality and size of the design are also verified.

5. *Layout:* Layout is the design step when a simulated schematic is delivered to the layout designers to generate the polygon-based representation of the circuit.

6. *Tape-out to manufacturing:* Once the layout of a full chip is finished and checked against process requirements and against the final schematic, it is the time to prepare the data for manufacturing.

4.3   ASSPs

An application-specific standard product (ASSP) is a standard product that has been designed to implement a specific function, as opposed to a general-purpose product such as a DRAM. In general, at first big companies had a monopoly over VLSI design, and most of them developed products that would sell in large quantities. These are standard products, not ASSPs.

Startup companies looking to leverage a certain area of expertise or define niche applications typically produce ASSPs. Today we have chips in almost every part of our lives: answering machines, car diagnostic computers, Global Positioning Systems (GPS), cellular phones, coffeemakers, and power supplies for various appliances. Each of these applications requires specific chips, and to be profitable in a competitive environment, the startup companies had to develop new methodologies that required new tools.

Designers of different ASSPs use slightly different design flows, because the ASSPs can be classified into different types as outlined in the following sections.

4.3.1 DSPs

Digital signal processing, or DSP, is carried out by digital circuits designed to address a broad class of problems in signal reception and analysis that have traditionally been solved using analog components. DSP is rapidly replacing analog signal processing functions where requirements for stability over time and temperature variations are critical. DSP is used to enhance, analyze, filter, modulate, or otherwise manipulate standard real-world functions, such as images, sounds, radar pulses, and other signals, by analyzing and transforming waveforms (e.g., transmitting data over phone lines via a modem).

Based on the complexity of the design or the cost of development, DSP designs can be implemented in an ASIC flow or in a flow that is similar to the microprocessor flow, in that it is primarily a full-custom flow.

4.3.2 ASICs

Application-specific integrated circuits (ASICs) are semiconductor circuits specifically designed to suit a customer's particular requirement, as opposed to DRAMs or microcontrollers, which are general-purpose parts.

The challenge of an ASIC flow is that typically the design is new and specialized, and there is no previous history on which to base architectural decisions. In this case there is significant emphasis on defining and verifying the architecture of the design.

Another characteristic of ASICs is that the designs are heavily biased toward logic structures. The design entry of this type of circuitry is tedious in a full-custom environment and not practical for designs exceeding 5,000 transistors. HDL methodologies therefore are standard for these designs where transistors are completely hidden from the designer.

In terms of layout, the picture is very similar. The layout designer is no longer involved in transistor-level design and architecture, but in block-level using advanced place-and-route tools. He may not actually see the full layout view of the library because for place-and-route purposes the obstruction shades are enough to generate complex blocks or chips. We will develop the library concept further in the next chapter.

Now, let's see how an ASIC flow works and what the various stages are to complete a design (Figure 4.2).

1. *Architectural/behavioral design:* See the definition of chip architecture in Section 4.2.

2. *RTL design:* Designers are developing and reviewing system-level and functional Register Transfer Level HDL code and implementing the desired functionality. Verilog and VHDL are the standard languages used for this function.

3. *Logic design:* Digital or functional simulations are performed as part of validating a behavioral model of the intended design. The simulation verifies that the chip architecture is feasible and will perform the desired operation.

4. *Logic/timing optimization:* This step is the most famous one and has revolutionized IC design from the days of full-custom schematic-based designs. The HDL code implemented in previous steps is useless without the ability to synthesize the code.

   In this stage, synthesis tools require two inputs: the design functionality in terms of RTL code, and a standard cell library with synthesis views and timing information. For each function coded within the HDL, the synthesis tool will chose the most appropriate library cell or combination of library cells to perform the job. The end result of synthesis is a netlist that contains standard cells and their connectivity.

5. *Place-and-route:* Place-and-route tools (P&R) are automated tools that require the following:

   - Standard cell library physical information, i.e., cell sizes, points of connectivity, timing, routing obstructions
   - A synthesized netlist that details the instances and connectivity relationships including constraints and critical paths in the design
   - All the process requirements for connectivity layers, including design rules of the routing layers, resistance and capacitance, power consumption, and electromigration rules for each layer

   Using this information, the layout is implemented automatically and optimized for minimum area and ideal timing.

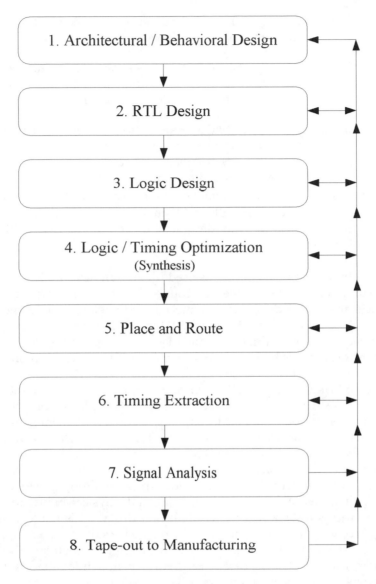

Figure 4.2 Detailed ASIC flow.

6. *Timing extraction:* This is a step to extract and calculate the timing of interconnect signals after the cells are placed and routed. Delay information is
   produced that is fed back to the circuit simulator for reverification of the
   design after layout. The result of extraction is a file in a format such as Standard Delay Format (SDF). This file will be used to go back and annotate the
   netlist with the real values from the physical design.

7. *Signal analysis:* Using this new netlist, the designer can now resimulate the
   design and find if the functionality and the timing specifications are met.
   In general, there are many cycles of P&R, extraction, resimulation, and
   synthesis until the specifications are met.

8. *Tape-out to manufacturing:* When the circuit specifications and layout design rules are met, the layout data goes to manufacturing.

This is the basic ASIC flow that has evolved over the past 15 years. As the complexity of designs grew, the tools and methodologies improved. Improvements to the basic layout flow include the following:

- *Increasing tool capacity:* One competition in the P&R marketplace was about who can place and route more gates in one run. When the size of the chips grew to more than 250,000 placed objects, it became inefficient to do the job flat, so the market moved to hierarchical design.
- *Introduction of floorplanning tools to the ASIC flow:* Hierarchical layout is accomplished using floor planning tools that were previously used by the full-custom chip designers to coordinate hierarchical requirements between blocks.

In order to select the cells used by the synthesis tools, we must make many assumptions related to the interconnect delays. The initial methodology was to use statistical models of interconnect loading before a place-and-route job.

When the timing delay was 100 ns, it was possible to make an error of ±10 ns because of the variation in the placement and routing tools and the expertise of the designer. When the processes got to 0.25 µm and the timing of critical signals dipped below 10 ns, nobody could afford a 10-ns mistake. Therefore, the solution was to bring some physical information into the ASIC flow sooner.

How can we bring layout data into the floorplan before the chip has real cells and real interconnect data? By running a global placement algorithm and a global router, we can extract from layout for top-level routing loading information for these signals. The data is not 100 percent accurate, but it is much closer to reality than the assumptions made before.

Without floorplanning and global routing it took five to ten iterations to get to a final layout. Using floor planning and physical information earlier reduces the process to two or three iterations per design. It is not a single-pass flow, but it is converging more quickly to the desired result.

As will be discussed in detail in Chapter 10, there are many links between tools. Especially in the case of P&R, it is recommended that the links between the floorplanner, placer, and router be very close. Tight integration of all of these tools is essential.

4.3.3 ASM

Application-specific memory (ASM) is a chip based on the ASIC flow, but containing a large memory block. Almost every chip today contains memory parts. They can be static RAM, dynamic RAM, ROM, etc., but in general they are no more than 512 kbits in size. An ASM is a chip that requires a lot of memory, from 1 Mbit up to 64 Mbits. Examples are video RAM (VRAM), synchronous graphic RAM (SGRAM), and in some cases more complicated designs.

The design of a memory block as described fully in the next section is done in a full-custom style. If memory expertise is part of the overall design process, then the chip could be implemented in a full-custom flow as well, or an embedded memory block (IP) could be the answer. Large ASMs would really be classified as a system-on-a-chip flow as described in Section 4.5.

4.4   MEMORIES

A very wide variety of memories are on the market today. Memories are a commodity part, as they are produced in high volumes for general-purpose consumption. A memory is conceptually simple: its job is to store data for retrieval at any time.

Examples of memory ICs are ROMs, dynamic and static RAMs, EPROMs and EEPROMs, SDRAMs, FRAMs, and SLDRAMs. The full list is quite long. Here is a short glossary of the different types:

- *ROM (read-only memory):* The contents of memory are fixed and can only be read.
- *SRAM (static random access memory):* Operates very quickly and remembers the data as long as power is supplied to the chip.
- *DRAM (dynamic random access memory):* Similar to an SRAM, but denser and needs to be refreshed periodically or else the data is lost. The data is dynamic.
- *PROM (programmable read only memory):* Generally programmed once and used as a write-once, read-many-times device. Can be electrically programmed and/or erased as well.
- *HDRAM^{TM} (high-density random access memory):* An embedded memory macrocell implemented in an ASIC process technology. The advantage of this is that a special manufacturing process is not needed. MOSAID Technologies Inc. has patented many of the techniques used for HDRAM.

In terms of design flow, memories are all very area intensive. Key metrics to memory effectiveness are the density and efficiency of the memory.

The relationship with the process specialists is very tight because memory processes are in general one generation ahead of others. They have to be ahead in terms of minimum manufacturing gate size, i.e., 0.25, 0.18, or 0.11 μm, because the memory chip depends heavily on process characteristics. If in a microprocessor we can get today 10 million transistors, in a DRAM memory the number has easily passed 256 Mbit for production.

Memories are implemented using full-custom techniques, and this fact shows the emphasis on area as the key issue. Every micron counts in memory design! Consider that current DRAM memories contain 256 Mbits per chip. Any reduction in the area of the memory cell reduces the chip's area by the same amount—multiplied by a factor of 256,000,000!

An obvious feature of a memory is that it is a very repeated structure with many cells that are pitch-limited in one direction. Therefore, in terms of design flow, the memories have a very interesting bottom-up and top-down design methodology. The chip and memory architectures are defined first, and then all layout starts from the memory cell and builds outward. The chip size is determined primarily by the memory cell size and the associated layout around the memory cell, so careful full-custom layout techniques are used.

4.5   SYSTEM ON A CHIP, OR SOC

The hottest design flow today (1999) is system on a chip, or SOC. This is the case when there is a need for a very big and complicated chip and there is no time to design everything from scratch.

The magic words in these types of chips are *design reuse* and *intellectual property* (IP) blocks. In design reuse, the individual blocks are design from the start with reuse in mind. The disadvantage to this methodology is that time to have a block designed for reuse is longer than a normal flow. The reasons are obvious:

- The specifications must cover common problems, not just specific ones for the current project
- More general-purpose simulations must be done
- If the blocks are implemented in a specific process, then the architecture will not be custom designed for it
- Additional consideration must be given to process variations

There are, however, big advantages when reusing the block as a core for future designs:

- Preservation of the IP—some experts may be busy on the next design (or have left the company)
- Time to design the second or third one is lower by up to 50 to 70 percent, depending on the kind of the imported block (hard, soft, etc.)
- The block or concept was verified in silicon, a fact that gives the next design team confidence that the block works

In terms of the entire design flow a SOC design will be a mix and match of different flows based on the type of each particular block. There might be control blocks done in an ASIC style and analog blocks or memories that require full-custom techniques. The integration of all of these block types is the real challenge of SOC design.

Let's see now how we can determine a flow that meets the requirements of SOC designs. The flow will depend on many factors and should be determined after many issues are understood. Here are some things to consider:

- *Technology selection:* What kind of process will be used? In the case of embedded memory, perhaps a blend process will do it.
- *Type of board on which the chip will be assembled:* From this it may be possible to determine the packaging type, footprint, die size, price, pin positions and assignment, and power consumption limitations.

- *Availability of core/blocks internally or on the external market sold by vendors:* Are they soft or hard cores?
- *Libraries available for the chosen process:* Are they silicon proven? What power consumption, speed, and tools are they compatible with?
- *Levels of testability to be addressed in the design:* This is another hot topic of the year—design for testability (DFT).
- *Limitations of the manufacturing process:* Time frame, special layers, reticle limitations, packaging limitations, etc.
- *Reliability of the chosen process:* Is it experimental, first-time trial, proven over a few working chips?

In terms of layout, the biggest job in these designs is to prepare a hierarchical design that has clear definitions of all the interfaces between blocks, whether internal or imported—floorplanning. Importing IP blocks can be tricky if the provider did not take into consideration any available standards such as those set by the Virtual Socket Initiative Alliance (VSIA), or if the process is not the same.

Blocks may be "soft" in that they are not process specific and need to be carefully implemented in the target process. In the case of "hard" blocks, the layout must be instantiated or migrated at the block level, extracted, and back annotated into the simulation to check timing and functionality in the new process.

No matter how up-to-date the flows presented here may be, by the time this book is on the market, design engineers, process specialists, and software developers will have found new problems. Remember that we, the user community, have interesting jobs in which we must use creative thinking to solve day-to-day design problems.

4.6   CAD TOOLS AS PART OF A FLOW

Each step in a flow is usually based around a specific CAD tool to perform the required operation. The choice or understanding of a tool within any flow depends on many factors, and it is these factors that we will discuss in this section.

The first concept to understand is that fundamentally all CAD tools fall into one of these two categories:

- *Design entry:* Methodology to implement the idea into a useable form with all the desired characteristics
- *Design validation:* Methodology to analyze and verify that the design has been entered correctly (i.e., it functions appropriately, performs as required, and is manufacturable)

Different tools address these two requirements in different ways or may address different issues. It is these different approaches that have spawned an entire industry in which each vendor tries to find the magic formula to develop the most effective design flow.

Over time, the number of tools to choose and understand has dramatically increased because there has been a tremendous growth in the variety of the following:

1. *Design types:* For example, the flow for a microprocessor design differs greatly from that for an analog component; therefore, a different set of tools is used. In this example there are design size and complexity differences to manage as well, which necessitates a different flow between them.

2. *Capture techniques:* Certain tools operate at different levels of abstraction in order to enable designers to capture their design ideas more efficiently. The difference between floor planners and polygon editors demonstrates this concept. They are both layout entry tools, but they capture design ideas at the block level (floor planners) or the transistor level (polygon editors).

   There are different capture techniques in the circuit entry domain as well. Schematic capture is one type; use of an HDL (high-level description language, usually VHDL or Verilog) is another. Schematic design captures a design at the transistor level; use of an HDL captures a design at the RTL (register transfer level).

   The difference between these two types of circuit entry has great impact on the layout design, as there are significant differences in the database format, size, and complexity of the resulting circuit design. An RTL-based design may result in anywhere up to millions of instances to layout. The flow for this design will be quite different from schematic-based flow, which produces much smaller designs.

3. *Design size and complexity:* One category of tools is one that automates tasks that would be logistically impossible for a designer to complete by manual techniques. Place-and-route tools that can implement millions of instances are examples of automation tools in this category.

4. *Degrees of specialization:* Each tool focuses on solving or addressing a small number of the required characteristics for a layout design. For example, the steps of routing signal lines within blocks and doing the same between blocks are often separate and use different tools, as they each have different requirements or constraints that cannot be covered by one tool.

   It is likely that there are separate tools available for each of the characteristics listed in Table 4.1, thus leading to a complicated flow for sophisticated designers who need to address all of these issues.

5. *Interface points:* There has been a significant trend toward having feedforward and feedback loops in the design flow. This has increased the number of interface points for each step in the flow. For example, floorplanners give interface information to other layout tools as well as to circuit verification tools. Layout editors must also be able to feed their results back to the floorplanning tool as well.

   Circuit design information should be flowing from the circuit designers forward to layout (as constraints or goals), and the results of layout at each step should be fed back to the circuit designers for verification. If this information is transferred more often and to greater levels of accuracy, then the overall design process will produce a much better result.

6. *Accuracy requirements:* Design tools have had to increase their abilities and accuracy to be able to implement new process technologies. Extraction tools have increased in accuracy, with examples such as 3D field solving techniques for near-body capacitance calculations.

7. *Acceleration techniques:* Often an existing step in the flow can be accelerated using a new tool that approaches the problem in a novel way. The functionality of the step is unchanged, but the algorithm in the tool accelerates the process. An example is the emergence of hierarchical layout verification that is many orders of magnitude faster than the previous flat hierarchy approach and provides identical functionality.

8. *Database formats:* Some companies develop proprietary database formats for their tools. These companies also develop tools of equivalent functionality to those already available in order to provide designers the complete capabilities they need using the proprietary database formats. Alternatively, they develop translators that interface other database formats to theirs.

   There are also issues with database compatibility for the case when circuitry to be implemented has come from a previous design or from a third-party supplier. In some cases, leveraging the experience of others through reuse or block-based design may affect the choice of tools in the flow. Typically, tools and/or steps in the flow are added to verify and translate the circuitry so that it is compatible with the current set of tools in the flow.

Using specific examples, the remainder of this chapter concentrates on the impact to the flow of the different design types that are common today. The type of design usually defines the tool set, mainly because the design size and complexity dictates the tool set that will accomplish the design in a reasonable amount of time. These examples will demonstrate the concepts we have presented in order to help you understand the appropriate set of tools for a specific design type.

4.6.1 Analog IC Design Flow

The greatest number of ICs shipped today fall into this category of products. The functions of analog ICs range from single transistor devices to complex functions that are characterized by very precise operating characteristics. Operational amplifiers, converters, and phase-lock loops are only a very small sample of these device types.

These devices are designed to exhibit extremely accurate analog characteristics. Output voltage levels and power consumption are very carefully controlled. Extremely detailed control of the manufacturing process is required because the circuitry on the chip is very sensitive to small variations of the transistor characteristics. Yield and reliability issues are examined more thoroughly than other device types because a large number of devices are produced. Table 4.2 summarizes key characteristics that are of most concern in the design of analog ICs.

TABLE 4.2 Critical Characteristics of Analog ICs

Behavior	Timing	Power	Area	Interface
Process	Maintainability	Reliability	Porosity	Flow compatibility
Complexity	Noise immunity	Yield	Electromigration	Latch-up immunity

Typically, understanding and capturing the fundamental behavior of the circuitry is straightforward. A small proportion of the circuit designer's time is spent to verify that the behavior of the design has been captured correctly. Compared to other design types, the number of transistors within the design is very small.

What type of flow is used to design this type of IC?

The answer is that it is probably one of the oldest design flows that exists unchanged today. At one time all IC design was done using the fundamental flow presented here.

This type of flow is generally known as a "full-custom" design flow (Figure 4.3), a term that comes from the freedom to completely customize all aspects of

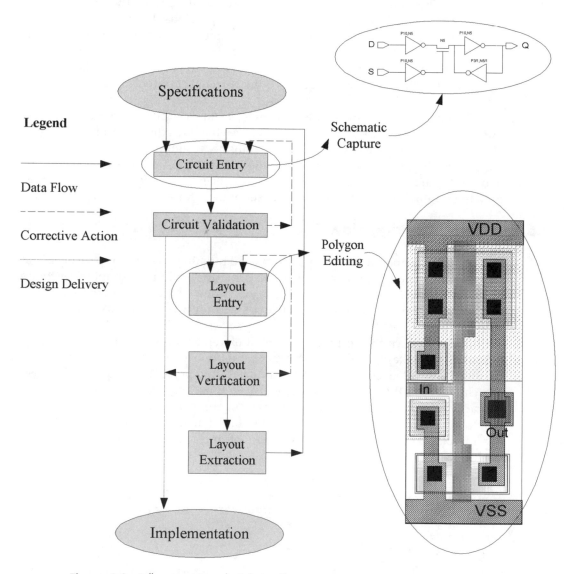

Figure 4.3 Full-custom (or analog) design flow.

the design. Transistor sizes and the layout implementation are both carefully designed to ensure that the final design will perform as intended.

The flow shown in Figure 4.3 is straightforward. From the specification, the circuit is entered and validated. Once complete, the layout is implemented, verified, and, as a final step, extracted and resimulated to ensure that the physical implementation operates as designed.

The circuit entry step is known as *schematic capture*, where transistors and customizable logic gates are entered as the implementation of the design. Components are manually selected, placed, and connected. The graphical look of the schematic is purely aesthetic from a flow point of view, but an understandable schematic is crucial for the design team to develop, communicate, and debug the circuit implementation of the design.

The layout is captured as polygons or paths. A *polygon editor* is used to capture and draw each individual transistor. Very detailed knowledge of the manufacturing process and the relationship between layers is required at this level of layout. It is also very important that the impact of different layout implementations on circuit performance be fully understood.

The schematic capture and polygon editing tools form the foundation of the full-custom design flow. These tools are primarily used in the role of *capturing a useable representation of the idea* as stated in our definition of a flow.

Automating this type of flow is difficult because the main issues that the design team needs to address are not easily solved using existing CAD tools. Issues such as noise immunity, reliability, and yield are left to the design team to evaluate once the layout is complete. The circuit validation process typically includes a large amount of visual inspection of waveform databases from detailed simulation models.

As such, the bulk of the analysis to determine whether or not the design *exhibits the appropriate characteristics for its intended use* is a labor-intensive affair requiring expertise in transistor operation and the physical characteristics of the manufacturing process.

In summary, analog circuit design demands the accuracy and control in implementation that a full-custom design flow provides. This is because of the design's requirement for very precise and stringent analog performance characteristics. Schematic capture and polygon editing are the fundamental tools of this flow.

4.6.2 ASIC Design Flow

In the IC design industry, it is common to hear people say "The heart of our system is a complex ASIC" or "Our design was implemented in an ASIC style." Companies advertise for people with "ASIC design skills." This really means the company is looking for people who are familiar with the ASIC design flow. It is the industry's definition of an ASIC design flow that will be described here.

In a strict sense, *ASIC* (application-specific integrated circuit) is a generic term that describes components that have been designed for a specific application and not as a multipurpose device.

For example, a telecommunications system could be designed using standard components such as counters, logic gates, and flip-flops. On the other hand, if an ASIC component was used, the counters and logic gates would be integrated on one IC and the function of the IC would only be useful in the system for which it was intended. In some cases these ASICs could be sold as a standard product to be used in multiple systems. This last example is normally referred to as an ASSP (application-specific standard product).

Theoretically, any ASIC could be implemented using the full-custom design flow described in the previous section (or any other flow, for that matter). In the IC design industry, however, the term ASIC has become far more synonymous with a certain design flow than a design type, although the term ASIC is used in both contexts.

What is meant by an ASIC design flow?

"Synthesis" and "Place-and-Route." These two very common terms in our industry capture the essence of an ASIC design flow. Their methodologies have revolutionized the way IC design is done today.

First let's understand the underlying principles of this flow, shown in Figure 4.4.

The following are some key points:

- *Circuit entry:* The design is implemented using a software language that is commonly known as RTL (register transfer level), but is in fact an HDL (high-level description language) written at the RTL level.

    VHDL or Verilog are examples of HDL languages that are used to capture design information. These languages support many different constructs. The most common way of describing an idea using an HDL is to write the code at a level of abstraction called RTL. At this level of abstraction, the code can be *automatically* converted to logic gates and sequential elements such as flip-flops and latches. This process is generically known as *synthesis*.

- *Layout entry:* The logic gates and sequential elements produced by synthesis tools are automatically placed and automatically connected using a P&R (place-and-route) tool.

    Note that P&R tools are designed to produce layout that is "correct by construction"; therefore, layout extraction for simulation is a step in the flow before layout verification. Layout verification is always required as a final step to ensure the integrity of the layout database and to check that any additions or changes made after P&R are verified.

- *Library of cells:* A prerequisite to this flow is the existence of a "library" of cells. The library consists of the logic gates and sequential elements that the synthesis and P&R tools use. In any design flow that follows an ASIC flow completely, the entire design is implemented using only the library cells and nothing else.

The ASIC design flow is probably the most common design flow these days, with the majority of flow and tool development supporting this methodology.

In an industry where the level of technology advances very quickly and products become out-of-date within a few years, the primary business focus of successful companies is to produce new products very quickly and productively. Minimizing the time-to-market of any product is crucial.

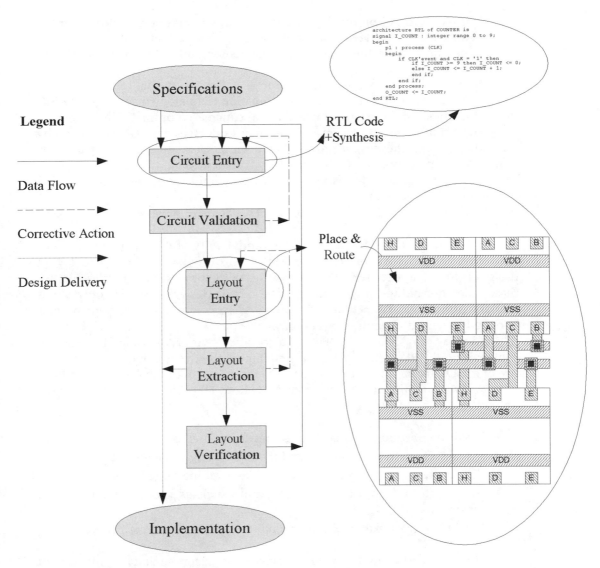

Figure 4.4 ASIC design flow.

In many ways the ASIC design flow has enabled this revolution in productivity, especially when compared to the older full-custom design flow.

In this flow, circuit entry is simply "coding" and therefore is very quick and easy. A large amount of functionality can be captured in a very short time. It is easy to maintain and reuse, and it is not dependent on a particular manufacturing technology. These are all significant benefits.

The example of code shown in Figure 4.4 is VHDL code of a counter from 0 to 9 (2 minutes to enter at 25 words per minute). Drawing a schematic is much more time consuming. Imagine changing the counter to decrement from 100 to 0 in a schematic-based design (or in a full-custom layout flow)!

This explosion of productivity on the circuit side has been matched for the layout-entry step by the development of P&R tools. Full-custom layout techniques are not practical when the circuit design is tens or hundreds of thousands of instances!

Advances in IC manufacturing processes have been one of the primary drivers of the development of all of the automation technology that forms the ASIC flow. Available transistors on a single chip have increased dramatically over time.

The ability to use this increasing number of devices has been addressed by this flow because the design process has been elevated to the logic-gate level, where logic functionality is the primary concern. The transistor level and many of the process-related issues are hidden within the library and modeled throughout the ASIC flow.

Table 4.3 summarizes key characteristics that are of most concern in an ASIC design flow.

Not only has this flow increased the productivity of IC designers, but it has also attracted many people to the world of IC design. The circuit and layout designers are separated from the complexities of the process design rules because these rules are hidden in the library. The circuit designer worries only about implementing logical functions. The layout designer may get by with process knowledge related to routing layers only.

To recap, an ASIC design flow is a specific design type using HDL coding, synthesis, and P&R methodologies to implement the design. In this case it is the use of a HDL for circuit entry that forms the foundation of the ASIC design flow.

We can infer the many tools of an ASIC flow. HDL circuit entry requires the following:

- Circuit verification methodologies specific to HDL code
- The use of synthesis to implement the HDL design to logic gates
- Circuit verification methodologies for the resulting large logic gate design
- Tools to feed forward circuit constraints such as groupings and timing constraints
- Tools to convert extracted layout data to relevant circuit verification data

The size and complexity of the resulting design mandates the use of different layout-related tools from those for a full-custom design flow:

- Floorplanning tools to group and guide placement of cells
- Automatic layout entry methods in the form of P&R
- High-capacity layout verification and extraction tools for large designs

TABLE 4.3 Critical Characteristics of an ASIC Design Flow

Behavior	Timing	Power	Area	Interface
Process	Maintainability	Reliability	Porosity	Flow compatibility
Complexity	Noise immunity	Yield	Electromigration	Latch-up immunity

There are additional tools and methodologies to generate libraries:

- Characterization tools to produce models for synthesis and simulation
- Implementation of cells in layout using a full-custom flow
- Alternatively, migration of cells from an existing library
- As another alternative, use of layout synthesis tools to generate library cells

In summary, an ASIC design flow is one that is most appropriate in implementing a complex and sizable logic design. HDL circuit entry, synthesis, and P&R tools are ideal for this type of design.

4.6.3 Memory IC Design Flows

Fast-page DRAM, cache, EDO DRAM, SRAM, and SDRAM are terms that should be familiar to anyone who has recently purchased a personal computer. The amount and type of memory in any personal computer is well advertised. We now understand that the more memory our computer has, the better and faster it will perform.

We are covering memory IC design in a separate section because it is a design type that is best implemented using a layout-first design flow. In terms of layout design, it is one of the few design flows where the layout is implemented before the circuit design!

This can be explained by first discussing the architecture of a memory IC. Figure 4.5 shows an example floorplan of a DRAM memory. Note that the majority of the chip area is consumed by the core memory cells and supporting circuitry.

The important concept to understand is that there are a relatively small number of leaf cells that are repeated literally millions of times. Any area savings that can be achieved for each leaf cell benefits the area of the chip many times over.

The most important example of this is the memory cell itself. Memory manufacturing processes are unique in that they have been specifically designed to achieve a small memory cell size. In the case of memory cell design, the unit of measure is more often nanometers rather than microns.

Figure 4.6 shows a design flow that is typically used in memory design.

The following are some key points:

- For memories, the key design characteristics are area, area, and area. Memory ICs are commodity products, and extra area equates to extra cost.

    The goal of a memory design is to pack a fixed number of memory cells in the smallest die. The design team's goal, therefore, is to minimize the area consumed by the periphery and nonmemory-cell circuitry. The ratio of memory cell area to the total die area is known as the *cell efficiency* of the design.
- As mentioned, the layout is done first. Typically, the memory cell is developed in conjunction with process development so the layout of the memory cell may be provided.

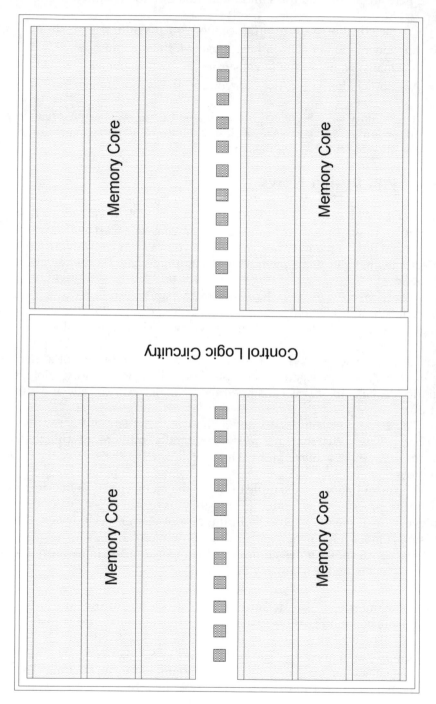

Figure 4.5 Example DRAM floorplan.

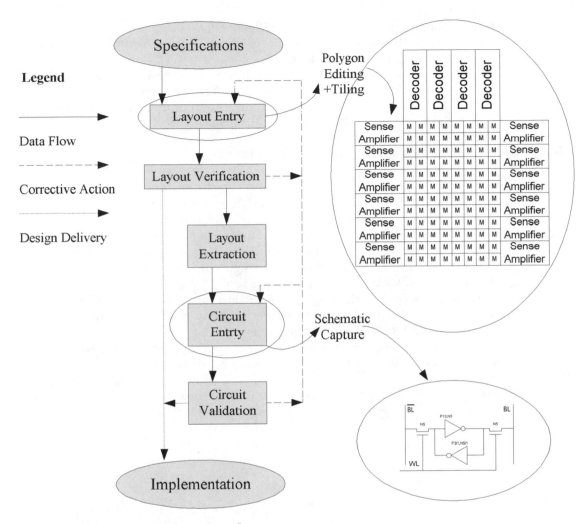

Figure 4.6 Memory IC design flow.

- This memory cell forms the basis for an inside-out layout flow. The memory core circuitry is built starting from the memory cell. The decoder and sense amplifier cells are "pitch-limited" in that the pitch of the memory cell is a limiting dimension for these cells.

- The design flow of the memory core pitch-limited circuitry is a very tight iteration loop between layout and circuit design. Many circuit design trade-offs are made to achieve a final implementation that is area efficient.

  These design trade-offs could be architectural as well. For example, the number of decoders and sense amplifiers is carefully chosen to ensure that a reasonable cell efficiency is achieved.

- Once the architecture of a memory design is well defined and validated, the architecture is often captured within a memory compiler or layout-tiling program for future use. Memories have a very regular structure and therefore are easily programmed into an automatic process.

- The control logic of a memory is still generally done in a full-custom style to minimize the total area of the die. Schematic capture is typically used as the design capture methodology.

    In summary, a memory IC design flow is one where the area is the first priority and therefore the layout characteristics are the key considerations during the chip's development. Thus, the design of memories is one of the more interesting challenges in layout design.

4.6.4 Microprocessor and SOC Design Flows

These designs have the most complex and intricate design flows and represent the state of the art in IC design. Microprocessor designs that are produced by companies such as Intel and Motorola combine all of the different flows that we have discussed so far. SOC (system on a chip) is a generic name for ICs that integrate a wide variety of complex and diverse functions onto a single die. In both cases it is the performance and cost benefits of a single chip solution to implement the required functionality that drives their development.

There is not one single design flow for the chip, but a mix of full-custom, ASIC, and memory design flows. There are variations of the three flows as well, depending on the block under development.

The effort and expertise that go into these types of designs is staggering and typically spans multiple years. Simply examining a floorplan of one of these ICs makes clear the variety and complexity of these designs. Each major block is planned and implemented using the flow most appropriate for its type. Once each block is complete, a separate flow of assembling the blocks is done.

CHAPTER FIVE

Advanced Techniques for Specialized Building-Block Layout Design

In this chapter we will start to apply our layout design knowledge to different types of layout leaf cells or "building blocks" that are common today. These different leaf cells are generally implemented in a full-custom design flow because they are optimized to form the foundation or library of cells that are used repeatedly to build an entire design.

5.1 STANDARD CELL LIBRARIES

A library of logic cells is the set of building blocks for the ASIC design flow discussed in Chapter 4. The library is typically called a "standard cell" library because of its common interface implementation and regular structure.

The library provides the functional building blocks used for synthesis and a layout representation of the cells for place-and-route. It is very important to note that the process of HDL synthesis limits the choice of logic cells to those that are found in the library provided. This guarantees that a physical or layout representation of the cells exists when the design is implemented using place-and-route tools.

5.1.1 A Brief History of Standard Cells

One way to understand the required layout characteristics of standard cells is to understand their history and the reasons behind their development. Once the concepts and methodologies behind this design process are understood, it is easier to fully appreciate the layout requirements for the cells themselves.

Why were libraries developed?

- Independent blocks became too big and complex for a full-custom design, so there was a need to speed up the circuit and layout design processes.
- There was a shortage of specialized personnel capable of hand-crafting complex full-custom designed blocks; automation alleviated this problem.

- Advances in the typical manufacturing process included increasing the number of routing layers from one to two or three metal layers. This added further complexity to the full-custom layout design process for optimal results.
- Even in a full-custom design flow, the placement of more than 20 cells is easier when the building-block cells are implemented with predefined standards. The standardization of cell interfaces is a concept that is implemented in a library.

The solution was to simplify the circuit and layout design of large digital circuits by using predefined and characterized "building blocks" (cells).

Before circuit synthesis tools were available, at first the idea was to develop predefined simple logic circuits. Examples would be inverters, NANDs, NORs, and flip-flops, to name a few. These would be designed and analyzed by an expert and then released to the project for general use. Everybody would be able to use these cells as building blocks for their circuits.

In order for a particular logic cell to be useful in different situations, the library was expanded to include each logic cell in a variety of sizes. Initially, the various sizes were decided haphazardly by individual designers.

A further refinement was to define the different sizes of each logic cell in the library so that the design can be more easily correct by construction. This is accomplished by following fanout guidelines in the case of amplifying a signal to drive bigger and bigger loads. For example, if the minimum size inverter is specified to be P2.5/N1.25, then the different inverter sizes in the library would be multiples of this size. A 2× inverter or INVx2 would have a size of P5/N2.5, INVx4 P10/5, etc.

Standardizing the sizes ensures that nobody will try to use weird numbers such as P8/N4 or P6/N3 in the circuit design phase, and maximum sharing of components is achieved.

When the synthesis flow was developed, the circuit designer actually didn't see the layout cells, so the need for standardization became even greater. As we mentioned previously, the synthesis tools automatically choose the best cell for the job; therefore, if a larger selection of cells is available, the synthesis tool has a better chance of optimizing the circuit.

Another factor that influenced the development of library cells was the impact of the first automatic place-and-route tools. The first automatic routers that came out started to change the way designers implemented full-custom connectivity, because the routing tools worked best with cells built in a certain way. Cell design was (and is) heavily influenced by the restrictions of the automatic tools.

Today the standard cell is the foundation of ASIC design. There are companies whose sole business is the design and migration of libraries into different manufacturing processes. Various EDA vendors provide circuit and physical design tools specifically for libraries as well.

ASIC design notwithstanding, the standard cell design methodology is also widely used to implement the "random logic" of a full-custom design. Initially, a circuit is partitioned into several smaller blocks, each of which is equivalent to some predefined function. Within each logic block, cells are implemented from a

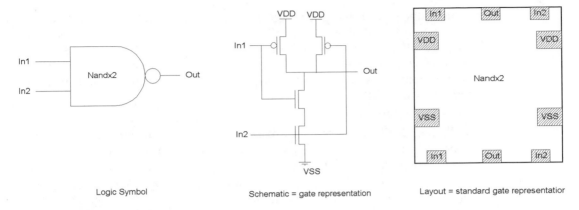

Logic Symbol   Schematic = gate representation   Layout = standard gate representation

Figure 5.1 NAND example of standard cell.

set of library cells. In general, the library is much smaller than a commercial ASIC library, but the methodology is the same.

5.1.2 Standard Cell Characteristics

A simple NAND gate is shown in Figure 5.1. A standard cell representation of the gate is shown. As we have discussed, an ASIC layout design is implemented at the cell or gate level, and detailed knowledge of the contents of the gate is not required.

In this section we will discuss the design of the cell itself. The goal is to give an understanding of the standard cell design so that it is compatible with an ASIC design flow. There are many issues to consider. First and foremost, however, is that the library of cells be compatible with the specific limitations or features of the manufacturing process to be used.

Typically, the design or architecture of the standard cells should be chosen based on the number of routing layers available in the target manufacturing technology. In certain special cases the design of the cells will depend on the characteristics of the available metal layers.

The following is a list of characteristics that are common to all standard cell libraries.

Characteristics related to the circuit design are as follows:

- The functionality and the electrical characteristics of each cell is tested, analyzed, and specified. In general, a test chip is manufactured and the performance of the each cell is analyzed from silicon. In some cases, only a process characterization step is completed to generate simulation models of the transistor characteristics, and library characterization tools use these models to create the simulation views of each cell.
- Multiple drive strengths for each cell type are created. In addition, the different drive strengths are multiples of a base or minimum size.

The following characteristics are related to the basic shape of the cells:

- During the layout design of the cells, the cells are built using a predefined template that will ensure that all the requirements are met. The template includes the height of the cell, the placement of wells, N transistors, and P transistors, and guidelines to follow so that the cell can be flipped vertically or horizontally and can be placed beside all other cells without creating errors such as DRC violations.
- Cells are rectangular.
- Cells for specific rows or chip areas are all the same height—a library may contain multiple sets of cells. For example, different cells will be used for logic, datapath, and I/O areas.
- Every cell length is rounded up to a multiple of a coarse grid. This grid is determined by either of the following:

  A specific design rule (such as the minimum well width)

  A desire to make placement easier and faster (using a coarse grid reduces the number of possible placement coordinates, thus accelerating the placement process)
- The power supply lines have a predefined width and position for the entire library—the width of the supply over the cell length is always consistent.

The following characteristics are related to the interface of the cells:

- All the input and output ports have a predefined type, layer, position, size, and interface points. These characteristics are determined based on the placer and/or router to be used to implement the design. The ports are targets for the router and should be optimized with the router in mind for best results.

An example of this would be that routing can be made faster and easier by using a signal pitch that is defined on a coarse grid. Routing tools will use fewer computing resources if a coarse grid is used because the arithmetic required of the tool is simplified.

- The interface of the cells can be designed to share certain connections. Examples would be source connections of transistors that are connected to power supplies. Alternatively, common substrate and tub contacts can be shared between cells.
- A rectangular outline and a set of obstructions for each routing layer are also characteristics of each cell. Obstructions can be defined separately for each routing layer, or the entire outline of the cell can be used as an obstruction. Obstructions can have any shape. They are not restricted to rectangles, but they have to be recognized by the routing tool.
- All nonshared polygons have to be spaced from the boundary of the cell by a value equal to one-half of the layer spacing design rule. This ensures that abutting cells will be correct by construction.

Other things to note about cell libraries:

- There are cells without any transistors, called feed-through or filler cells, that can be added between cells to allow vertical connectivity when there are no more routing resources over the cell.

- In the case of I/O cells, there are specially shaped cells for the corners of the chip where two rows of cells meet.

A typical standard cell library consists of hundreds of cells. Advanced libraries consist of more than 1,000 cells. There are cell libraries specially designed and developed for low power consumption, high speed, very good porosity, etc.

5.1.3 Standard Cell Architectures

As we stated in the previous section, the design and architecture of a standard cell is dependent on the number of routing layers that are available in the manufacturing technology. Let's investigate why. An example of a standard cell is shown in Figure 5.2.

Note that only one layer is used to make most internal connections to the transistors within the cell. This version of a standard cell is compatible with the early versions of channel routers. These routers could only make connection to cell pins that were placed on the boundary of the cell.

As you can see in Figure 5.2 the connections to the second layer of metal is made to the cell to ports only on the top and bottom sides.

The design shown in Figure 5.3 consists of two rows and a routing channel in between. Note the feed-throughs in the design. They are empty cells that are

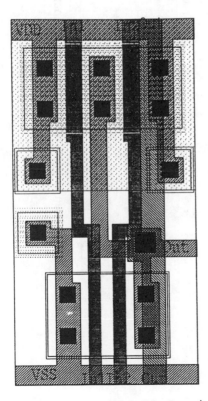
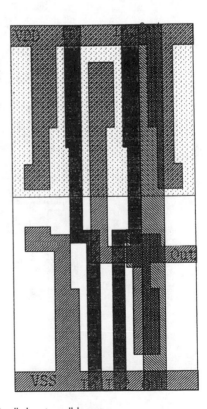

Figure 5.2 Example standard cell showing all layers.

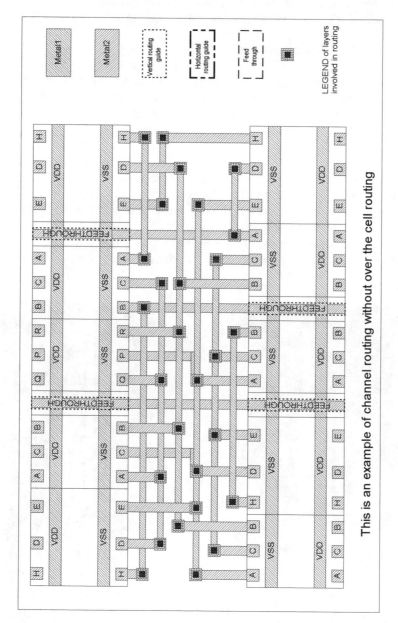

This is an example of channel routing without over the cell routing

Figure 5.3 Channel routing.

there to accommodate vertical tracks that connect signals located in different routing channels.

In a process with only two routing layers, feed-throughs are the only way to add vertical routing tracks. In this case the routing is done in two steps. First, enough feed-throughs are assigned to implement the required number of vertical running signals. The completion of the rest of the routing follows feed-through placement.

When a different design is done that accommodates vias in the middle of the cell, then ports can be placed there and over-the-cell routing is possible. Refer to Figure 5.3 for a clear example. In this case vertical routing tracks are possible over the cells and the use of feed-throughs is reduced in the final design.

Figure 5.4 compares the two cell architectures and the differences in the ports and interfaces to the designs.

Note that in Example A, the ports to the design have to be accessible on both the top and bottom sides of the design. This is to ensure that routing channels are used most efficiently and it minimizes the use of separate feed-throughs. The overhead of routing these ports within the cell to two sides of the cell is significant, adding parasitic loading to the signals and reducing the overall porosity of the final design. Examples B and C do not suffer from this overhead, and a cell equivalent to Example A may be smaller overall.

Example C in Figure 5.4 shows an advantage of over-the-cell routing and clever interface design. In this case, the connections to the IN1 and IN2 ports may be extremely short if the cell that is connected to it is placed directly adjacent to the one shown. In this case, space is not consumed in any routing channel.

Figure 5.5 compares two routed designs with and without over-the-cell routing. In both cases there are only two layers of metal available for routing, but the difference is whether or not over-the-cell routing is possible.

In the top picture, we can observe that the channel used for routing can only be the empty space between cell boundaries. In the second example this restriction is not as acute. The area of the second design is significantly smaller and for large designs the difference is amplified. The main reasons for the smaller area in the case of over-the-cell routing can be summarized as follows:

- The router can route horizontally between adjacent out–in port connections. This is a major source of area savings, as the routing channels between cells are reduced by this effect alone.

- The available channel size includes the cell area and therefore is more efficient.

- Porosity is much greater because of the metal1 horizontal connections between adjacent cells, because of the reduced cell size due to the elimination of routing port signals between top and bottom sides, and finally because of the elimination of feed-throughs.

In a three-metal-layer process, almost all the channels can be removed and all routing can be completed over the cells.

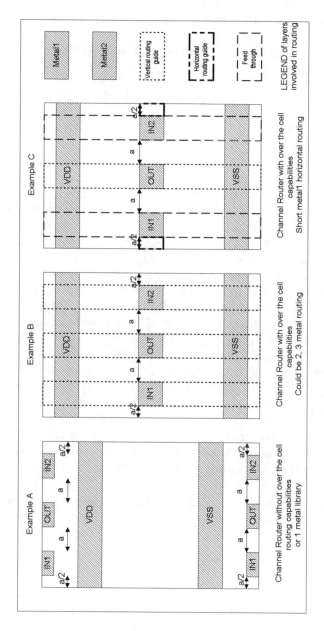

Figure 5.4 Comparison of cell interfaces.

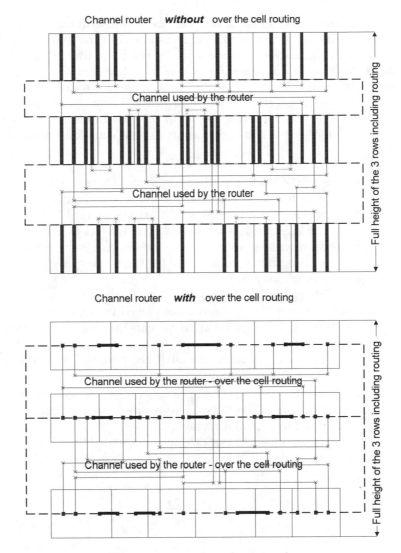

Figure 5.5 Comparison of routing styles.

5.1.4 More Standard Cell Concepts

We have mentioned the term "porosity" of a design. The porosity of a design can be used as a metric of quality and/or efficiency of a design. Porosity as it relates to layout design is defined as the ratio of the total available routing area to the total cell area. Cell pins are not included as routing area.

For example, in the case where many pure routing channels are used, the porosity is low. Ideally, we want to increase the porosity of a design by eliminating any channels and have only the logic cells that are needed to implement the design.

We may need to trade off porosity within a library to ensure that we can complete the design in a reasonable amount of time and take advantage of any features of the place-and-route tools. At the same time we need to avoid limitations that these tools might have. It is useful to read the manuals for the place-and-route tool to be used before designing the library of cells.

Among other things, we have already mentioned that a coarse routing and cell size grid is useful for reducing the compute resources of the place-and-route tools. This and other points to note when designing standard cells are as follows:

- The choice of routing grid is important. There are three types of grid used today: line to line, line to via, and via to via. Figure 5.6 shows examples of the three cases and how they may affect standard cell design.

  The line-to-line pitch suffers from the fact that a lot of signal jogging is required whenever a via is used. The cell size is the smallest, but routability suffers. Also, the execution time for the routing engine to complete the job will be much longer.

  Similarly, in the case of via-to-line pitch, signal jogging will occur whenever two vias happen to be placed on adjacent lines. Trying to avoid this case is an overhead that may negate the area benefit of the reduced signal pitch.

  In the case of via-to-via pitch, the porosity of the design is maximized and the routing is more straightforward.

- The standard cell height and maximum cell length for the library must be determined based on the tools to be used, the purpose of the library, porosity, and number of routing layers.

  For example, in the case where metal1 is the layer used internally to the cells, there is a rule of thumb that says, "The length of the longest cell

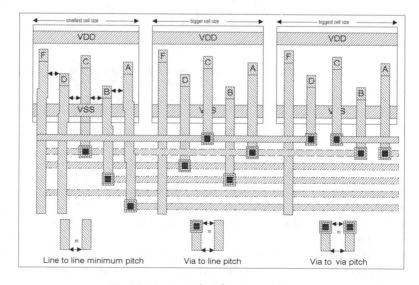

Figure 5.6   Examples of routing grid types.

has to be no bigger than six times the cell height." In the case where cells will be longer, the cell height should be dramatically increased (perhaps doubled). The reason for this is that the placement tools can work much more easily with a reasonable height-to-length ratio. A long cell is much more likely to have internal metal2 jumpers, so the porosity is reduced for routing.

- Power supply issues for the entire block must be addressed before the place-and-route takes place. Strapping and connecting the power supplies is always attacked first when implementing a standard cell design. The following issues should be considered:
  Connectivity between standard cells
  Tracks routed internally to the standard cells
  Electromigration applied to track width and number of vias
  Resistance

- Since the place-and-route tools are only concerned with the placement and connectivity to standard cells it is useful to consider the design of the cell abstract. The cell abstract is not a physical part of the design, but is used only by the place-and-route tool. The abstract consists of the following:
  Cell boundary
  Location and shape of target pins
  Routing obstructions

Routing obstructions or "keep outs" are areas within a cell that are unavailable for routing in the assigned layer. For example, very little area within most standard cells is available for metal1 routing because there is a lot of metal1 used for the local interconnections.

If the router is free to use metal1 in trying to connect cells, it has to know where it can and cannot place signal tracks. Obstructions within the cells give the router this information in the form of shapes on the given layer.

Figure 5.7A shows an example of a standard cell and the areas of metal1 use. One approach to defining the obstruction areas would be to replicate all metal1 shapes.

Even for small blocks (500 cells) there is a strong dependence between the compute time of a router and the way the obstructions are defined. The routing software algorithms needs to build a map of the routing porosity for every layer, so the more complicated the shapes are in the keepout layers, the bigger the files. With complex polygons the computational requirements become longer and more complicated and the routing time starts to grow exponentially. So let's see how we can help the tool to run faster and more efficiently, using fewer computer resources. Refer to Figure 5.7A for an example.

In Figure 5.7A you can easily see that the smart keepout is only one polygon on metal1 and that it has a minimal number of coordinates. Not only is the obstruction concisely defined, but there is an open space for the router to use in case there is a need to put down vias for routing.

To automatically obtain a shape like the one on the right we can use a simple macro that will do the following:

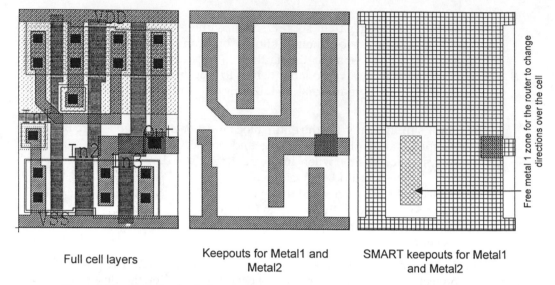

Full cell layers   Keepouts for Metal1 and   SMART keepouts for Metal1
                               Metal2   and Metal2

Figure 5.7A Example of cell obstructions.

- Oversize the polygons on each routing layer
- Merge all the polygons
- Undersize the polygons by the same amount

For example, to determine the amount needed to over/undersize the polygons, review the following calculation:

$$\text{Metal1 Width} = 0.5\,\mu\text{m}$$

$$\text{Metal1 Space} = 0.4\,\mu\text{m}$$

$$\text{Drawing Grid} = 0.05\,\mu\text{m}$$

$$\text{Oversize Value} = \text{Metal1 Space} + \frac{\text{Metal1 Width}}{2} - \text{Drawing Grid}$$

$$= 0.4\mu\text{m} + \frac{0.5\mu\text{m}}{2} - 0.05\mu\text{m}$$

$$= 0.6\mu\text{m}$$

Figure 5.7B shows a graphical implementation of the above steps.

The intent is to obtain one single shape that will cover all the small polygons. This shape will include the places where the distance between two of them is smaller than the possibility of routing another nonrelated line. Doing so for each cell in the library, for each routing layer, we can significantly reduce the place-and-route time.

Legal distance between metals for a feed-through signal

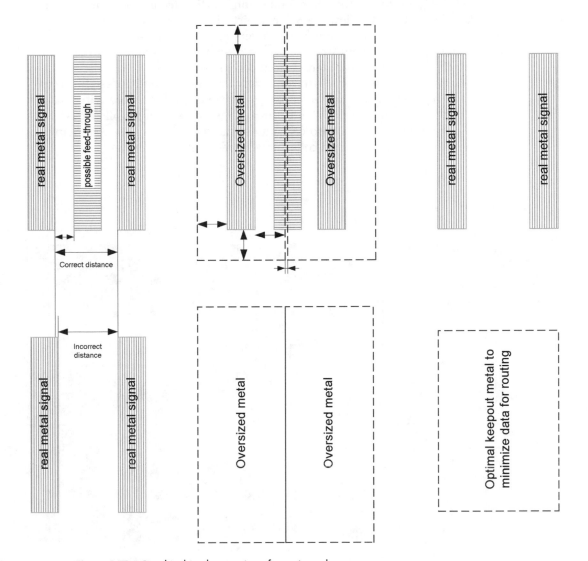

Figure 5.7B   Graphical implementation of oversize value.

5.1.5 Gate Arrays

This design style is a different approach to standard cell design.

Unlike standard cell design, where all layers are different from one cell to another, every gate array cell within one particular library has an identical arrangement of a base set of layers.

Typically, this base set of layers includes only the layers necessary for the formation of transistors; therefore, the list of base layers would include wells, active regions, implant layers, gate polysilicon, and nothing more.

Upon this base set of layers, individual "standard cell" functionality is obtained by patterning the interconnection of the transistors into different configurations. Refer to Figure 5.8 for a pictorial explanation of how cells are defined. Typically, the interconnection layers are the first one or two layers of metal. These defining layers are collectively known as the "shades" of the cell and are the only layers that are different from cell to cell.

The term *gate array* comes from the way in which an entire design is implemented. In a gate array design, an entire wafer can be prefabricated with an array of identically patterned transistors or groupings of transistors called base cells. Logical definition of the standard cells and of the connections between them uses only the shades of the required cells and the free routing layers.

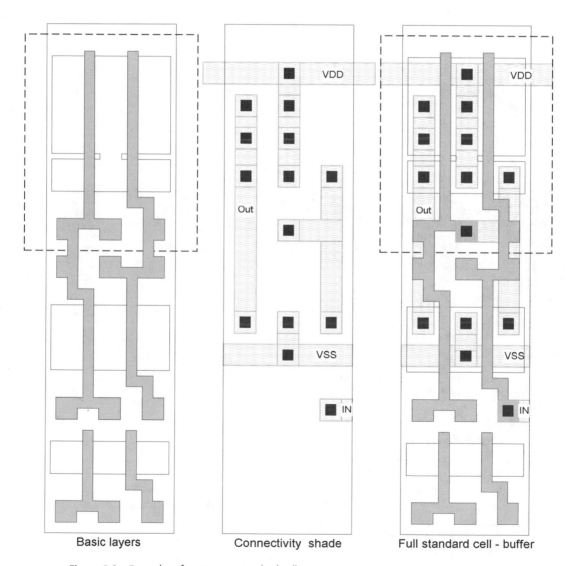

Basic layers	Connectivity shade	Full standard cell - buffer

Figure 5.8 Examples of gate array standard cells.

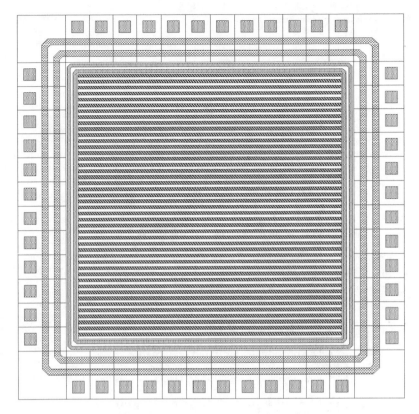

Figure 5.9 Example gate array master slice.

A wafer of base cells without any standard cell definition is typically referred to as a "master slice," as shown in Figure 5.9. Note that a master slice can be pre-fabricated while the circuit design is in progress or in fact manufactured in advance.

Table 5.1 outlines the trade-offs of gate array design to a full layer standard cell design style.

The simplicity of gate array design is a trade-off between the cost of rigidity imposed upon the circuit both by the prefabricated wafers and the advantage of the speed in the manufacturing process. Only the last few steps in the fabrication process depend on the application for which the design will be used. Thus, gate arrays are cheaper and easier to produce than full-custom or standard cell designs.

The idea of having master slices on the shelf that are ready for quick implementation can be very useful. Differences in circuit design size or pin count can be quickly addressed by having a select range of master slice configurations available for use.

Another scenario where special master slices are valuable might be where a single design idea has multiple minor variations on a theme. In this case, using an efficient gate-array master-slice approach may be viable.

TABLE 5.1 Key Issues in Gate Array Design

Benefits	Comparison to Full Layer Standard Cell Design
Faster design time	• Wafers are prefabricated with base layers
	• Iterations and optimization to cell placement and interconnect routing is greatly reduced because of limited placement sites and lower cell density
	• Revisions to existing design starts with prefabricated wafer
	• Many issues such as power supply routing and latch-up protection are addressed in advance, since the base layers and architecture of the chip have been predetermined. At the very least, the effects of these issues are much more predictable than the case where all layers are implemented at once.
Performance	• Better performance than programmable IC alternatives because there are more opportunities for layout optimizations
Increased reuse	• Master slices can be used for different designs or applications
	• Cell design easier to port to different process
Drawbacks	**Comparison to Full Layer Standard Cell Design**
Design flexibility	• Fixed availability of total number of gates
	• Less opportunity for layout optimization
	• Less opportunity for custom blocks such as memories
Cell density	• Generic base layer cell design results in overhead of unused transistors

Note that while custom blocks or "hard macros" are not typically used in most gate array applications, they can still be implemented to create special master slices. For example, if multiple applications or designs have a common requirement such as a similar amount of memory, then a master slice that incorporates the common requirements can go a long way in saving wafer and design costs. Figure 5.10 shows how a custom-made block might look in a gate array design.

Note that the reverse is also possible. In a full-custom design it is possible to embed a block of gate array logic! See the CD-ROM for additional examples.

When else might a gate array design style be useful? Scenarios where a gate array design style may prove to be useful and valuable include the following:

- The standard cell portion of the design is very risky or complex
- The standard cell portion of the design is behind schedule
- Both of the above are true

Consider the preceding scenarios where a special master slice is being created at the same time as the initial design is being implemented. The logic or standard cell portion of the design is on the critical path of the project. If a gate array design style is used, the manufacture of the base layers can begin before the complete design is finalized. Typically, the elapsed time to manufacture the base

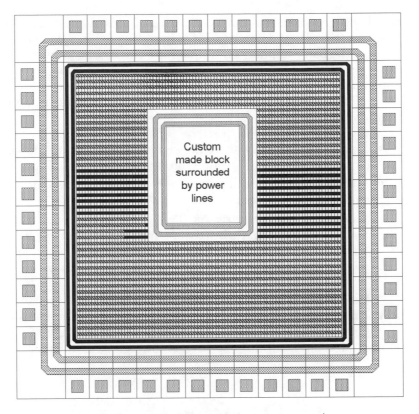

Figure 5.10 Embedded custom block in a gate array design.

layers of today's processes might be a full month. During this month the design team has an opportunity to catch up and finalize the last design issues. The alternative in a full layer standard cell environment is that the entire manufacturing process is delayed for a month while the design is finalized.

Please check the CD-ROM for pictures of a gate array that MOSAID designed some time ago. In one case the chip had several special custom blocks as part of the design. In another case the gate array logic was embedded in a full-custom design. The latter was a design by MOSAID for an Ottawa-based company called Accelerix.

5.2   SPECIAL LOGIC CELLS

5.2.1 Datapath Library Cells

The easiest way to understand what is datapath functionality is to consider the operation of an example circuit such as an arithmetic logic unit (ALU). The ALU is one of the three essential components of a microprocessor, the other two being data registers and the control circuitry. The ALU performs addition and subtraction, logic operations, masking, and shifting (multiplication and division) on multiple bits (signals) of data. An ALU is implemented in a datapath style.

As the name implies, a multiplier is a circuit whose output state is the arithmetic product of two input signals. This is another example of a circuit that is implemented in a datapath style.

What are the characteristics of datapath cells that distinguish them from standard cells?

- *Signal flow:* Typically, there are signals or bits of data flowing through the circuit, as you might imagine through an ALU or multiplier.
- *Multiple signals:* Several groups or buses of signals are flowing through the circuit at the same time.
- *Requirement for symmetry:* As the signals race through the datapath, it is highly desirable that each signal path be topologically identical to the others. This ensures that mismatches in timing do not occur and that the predictability of each signal relative to each other is known.

How do we attack these special requirements in a systematic and efficient way? The answer is to use datapath library cells and techniques.

Let's consider an example design as shown in Figure 5.11. This example shows four 8-bit signals being processes through three different functions (labeled

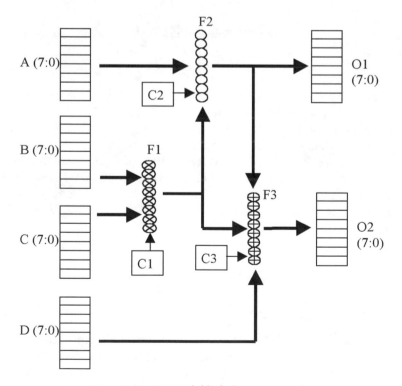

Figure 5.11   Datapath block diagram example.

F1, F2, F3) and producing two output buses O1 and O2. The control blocks C1, C2, and C3 control all 8 bits of the functions simultaneously.

One alternative in implementing this in layout is to floorplan the circuitry as shown in Figure 5.11. In this case the requirements of symmetrical signal performance between bits would be difficult to achieve.

Let's apply proper datapath techniques.

The first and most important detail is that the layout cannot start without a full picture of the functionality of the entire block. The way the signal lines have to run over the bits, their number and position, and the number of vertical tracks needed for internal cell connectivity are details that depend 100 percent on the schematic connectivity and performance requirements. In this case the design engineer and the layout designer have to work closely; otherwise, the full-custom block will not meet all design requirements.

The first step in proper datapath design is to consider the circuitry on a bit-by-bit basis. This is shown in Figure 5.12. Note that this should apply to both the circuit and layout design processes.

Remembering that the control signals are common to all 8 bits simultaneously, we can now start to see how we might achieve symmetry across all 8 bits.

1. Divide the complete functionality into smaller cells. In this case F1, F2, and F3 should all be separate cells. Each function is implemented as a row in the floorplan.

2. Define the interface to these cells first without completing the internal layout. This must include the internal and external routing requirements for the cells. Consider the number and layer for both the vertical and horizontal tracks and share routing channels wherever possible.

3. Fill in the internal layout of the cells.

4. Complete the entire datapath for one bit. It is a simple process to step and repeat the layout for the rest of the bits. The cells should be designed to abut to themselves in the direction of the multiple bits.

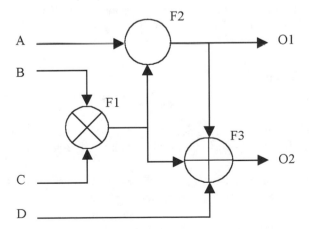

Figure 5.12 Single-bit data flow schematic.

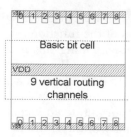

Figure 5.13 Example of datapath library cell.

Note that the schematic hierarchy and layout hierarchy should match for simpler verification and extraction. In our example, the floorplan of the layout will look similar to the flow shown in Figure 5.12 with the three functions implemented as cells and placed in the order shown.

Shared circuitry across all bits, such as the control circuits, should be considered in determining the cell breakdown and included in the floorplan.

Figure 5.13 shows an example interface design for one function for one bit of the datapath. This is considered to be one of the datapath library cells.

We can observe from Figure 5.13 that a datapath cell has many interesting features:

- The vertical cell interface starts and finishes with VSS power lines to allow abutment
- There are predefined vertical routing tracks for interconnectivity between the rows (functions)
- Signals can bend or jog over the cell when passing from one function to another as long as the signal exits on a predefined routing track

Figure 5.14 shows a complete implementation of a datapath circuit. Looking at the basic cell and the array of 3 functions × 3 bits, we can observe the following:

- Internal cell routing is shown and is done based on available tracks. Signals run freely and vias are placed centered to the lines or offset from center.
- As the picture shows, there are many unused lines connecting different rows. This was done intentionally to allow enough spare tracks for internal connectivity of the cells.
- All the routing is done for 1 bit only, and then the routing cell is arrayed over the width of the datapath.
- The rule of signal direction is respected here, too: M1 and M3 run only horizontally, M2 and M4 only vertically.
- All the gates generating the control signals are placed at the end of the row for each function. Routing tracks and efficient use of layout area are achieved this way.
- The end cells close off the substrate connections and overlap rules (for example, for N-well guard rings). Logic can be placed at both ends as required.

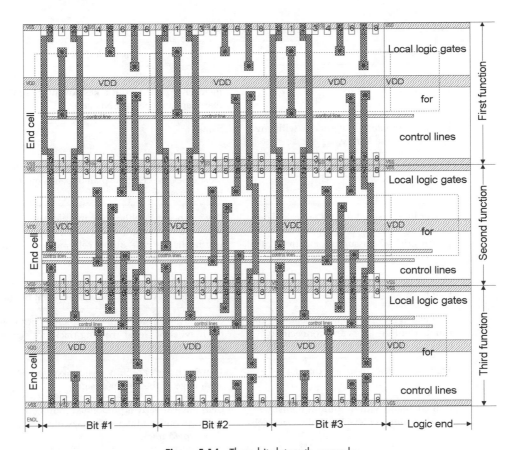

Figure 5.14 Three-bit datapath example.

When implementing the layout design for the cells, it is useful to understand the differences between datapath and the standard cell library cells (Table 5.2).

Because of the complexity of this layout, the color CD-ROM version of these diagrams may be easier to analyze.

A few comments about automatically place-and-routed datapath blocks. There are two kinds of datapath automation. One is a simple P&R of standard cells using a normal library and timing-driven limitations. The more advanced way to generate datapath blocks is using a specially built library with all the cells obeying the rules explained earlier. There are datapath-specific place-and-route tools that specialize in "bit" placement and routing and total symmetry of the signals in the bus.

5.2.2 Clock Generator Cells

Clock generator or perhaps clock buffer cells are generally special cells in a library. These cells are used specifically to buffer or amplify a very heavily loaded signal—namely, a system or chip clock signal. The system clock needs to be distributed all around the chip with as little delay as possible, and thus the implementation

TABLE 5.2 Standard and Datapath Cell Comparison

Standard Cells	Datapath Cells
Designed to interface horizontally with any other cell in the library.	Designed to interface horizontally only with themselves. The layout takes advantage of the fact that there are special END cells.
Designed to interface vertically with any other cell in the library.	Same
In general made of one N+ and one P+ region design.	Not defined or restricted to anything specific. In most cases the cell floorplan depends on the transistor sizes.
Power lines inside the cells are calculated for average power consumption and power grid—the assumption is that not all the cells in a row are working at the same time.	Power lines inside the cells are calculated to sustain a defined row length as determined by the bit width—in this case, all the cells are working in synchronization for all the bits in a bus.
Cell height is fixed.	Cell height is variable and independent of other cells
Cell width is variable and depends on the number of transistors and connectivity.	Cell width is fixed by the floorplan, which defines the number of "legal" routing tracks for signals over the cell.
Transistor sizes are standardized.	Transistor sizes are fully customized to take advantage of the layout design.
Cell are made with maximum porosity in mind—minimal metal2, metal3 inside cells.	Cells have already a predefined routing over the cell so the layout can take advantage of the known free or occupied tracks.
Each cell has only one standard function—for example, INV, FLIP_FLOP.	Cells may be designed to have multiple functions as defined by the over-the-cell routing (similar idea to gate array cells).
The entire placed and routed area has to be extracted to obtain timing information.	It is enough to extract only 1 bit to understand the timing for a datapath.

of the clock distribution system is a topic for layout in itself. In this section we will concentrate on the buffer cells.

What is so special about clock generator cells? The transistor sizes can be immense: 1,500 to 2,000 µm for a single device width is not uncommon. Compare these values to a minimum size inverter that can be 1.5 µm or so in width. To be implemented effectively, these huge transistors merit special layout techniques. The main concerns with such devices include the following:

- Optimizing signal and power connections in terms of resistance and capacitance.
- Healthy substrate connections—remember also that the clock in a chip is generally one of the highest speed signals and may generate a significant

amount of noise and coupling into the substrate. Generally these transistors are isolated with independent guard rings.

- Techniques to reduce supply resistance include busing wide connections from power supply pads and a large number of vias.

- Electromigration rules must be strictly followed.

- The timing characteristics of clock signals are critical, so extraction and simulation of the layout is a must.

- If there are different clocks that need to be synchronized, then the layout should be symmetrical between them. A common technique is to use one clock cell that is configurable as it is used in different locations.

Please refer to Chapter 7 for a detailed discussion of the layout techniques for larger transistors, as well as to the pictures on the CD-ROM for a clock generator cell example. The picture is simply too complicated to show in black and white.

5.2.3 Bus Interface Unit (BIU) Cells or a Barrel Shifter

In complicated chips, there are many buses of signals that are helping the various blocks to interface in, within, or with the external world. Nevertheless, many buses and many signals are taking precious space from the chip area. Various blocks are working based on different clocks, so the solution was to develop switches for these buses in such a way that they can be used at different times by different blocks.

A *bus interface unit*, or BIU, is one of the solutions to control the traffic of signals over one bus at different times. Different signals are essentially multiplexed onto a common bus at different times. As such, these bus lines can be heavily loaded with mixing circuitry, and it is highly desirable to minimize the chip area consumed by purely routing channels.

There are some challenges in meeting these requirements:

- Minimizing the bus capacitance
- Minimizing the area consumed by the bus itself and the muxing circuitry
- Achieving symmetrical and predictable performance for all connections to the bus

An approach that minimizes signal capacitance and maximizes the use of the routing area is to plan for and implement this specialized circuitry directly under the signal routing. Parasitic loads for each of the numerous connections to the bus are minimized in this approach, and if a common cell is used, then all connections to the bus will be similar.

In terms of layout design, layout implemented directly under a signal bus is one of the most challenging tasks. Only experienced designers in full-custom layout are able to achieve quality results. All aspects of layout design must be considered and juggled when implementing this type of design—for example, area, size, positioning, capacitance, resistance, symmetry, and layers to be used for routing.

The CD-ROMshows an example of this type of circuitry, as it is too complex for the final design features to be effectively communicated in black and white.

5.3   PAD CELLS

Every chip has an interface to the external world of the printed circuit board. The way an integrated circuit does this is through the pins of its package. These pins are connected inside the chip package to metal conductors that are collectively called a lead frame. The final connection to the chip is from the lead frame through gold bond wires to large metal areas that are called pads.

Pad cells are the layout cells that have the large metal areas or bond pads within them. It is natural and more reliable to make these layout structures separate cells to ensure consistency in characteristics for all pads. Pad cells usually incorporate several structures and are designed to provide the following:

- Reliable connection area for wire bonding
- ESD protection structure
- Interface to internal circuitry
- Optionally, the logic directly related to the function of the pad such as input or output buffers

Relatively speaking, bonding pads are very big, around $85 \times 85 \mu m$, because the pad is a target for a mechanical machine that is physically soldering a gold strand of wire. The pad metal is the top layer of metal and typically has a very large via connecting the top layer of metal to the underlying layers.

Figure 5.15 shows the pad structure and cross-section to show the combination of the metals and passivation or overglass layer. Note that the passivation layer is a negative layer. By necessity, the pad is not covered by the protective passivation layer (a glass material that protects the die) and thus is an area that is vulnerable to dirt and other foreign objects. Also note that the two metal layers are connected through a large via. For a greater variety of via connections, please refer to subsequent chapters.

We can observe that the no-polygon zone refers to a distance that is measured all the way around the pad generating invisible arcs. This feature can be used for passing 45-degree lines in between the pads.

In most ASIC designs, the pads are placed in a ring surrounding the core logic of the chip so the pad cells are designed using special geographical rules for layout. This is generally known as a *pad frame*. As there are three main types of pad cells—power, input, and output—the pad frame is also known as the I/O region. Examples are shown in Figures 5.16 and 5.17.

Figure 5.16 shows staggered pad cells. This arrangement accommodates more pads, since the pads can be placed closer together. In this case, the manufacturer requires a different and more specialized bonding machine. The increase in bonding complexity and cost is offset by the ability to accommodate more pads and by having a smaller and more efficient die. Since the bond pads have to be a specific physical size, in some cases the size of the design may be determined by

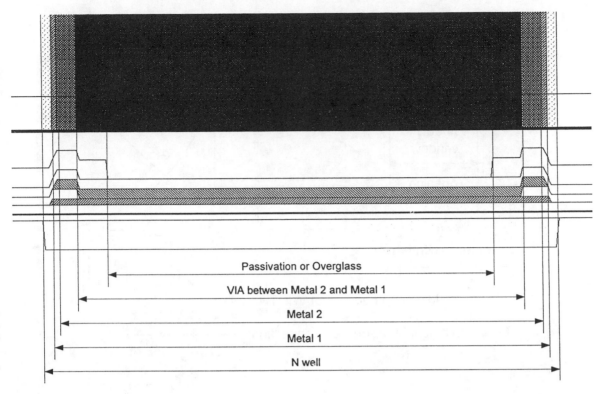

Passivation or Overglass

VIA between Metal 2 and Metal 1

Metal 2

Metal 1

N well

Figure 5.15 Bonding pad layout.

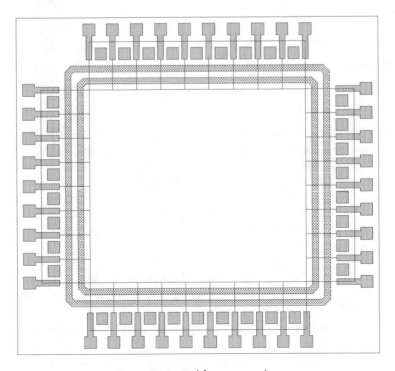

Figure 5.16 Pad frame example.

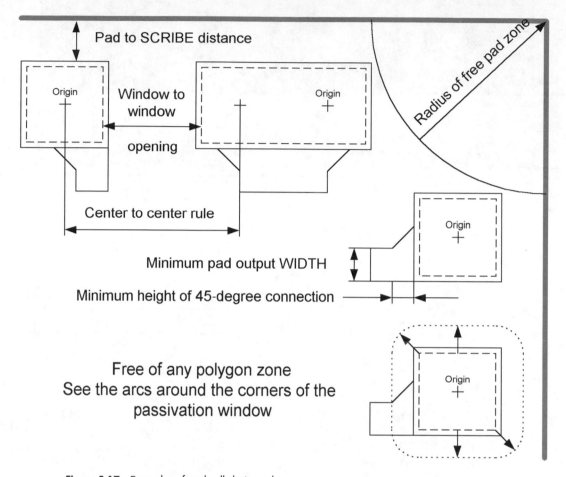

Figure 5.17 Examples of pad cell design rules.

the number of pads when the ratio of pad cells relative to the amount of logic is exceedingly high. It is desirable to avoid this situation.

Note that only the metal of the pad is in two rows. The power supply rings and transistors related to the I/O circuitry are still in one row. This is for consistency in circuit performance and ease of layout design and integration.

Let us now consider specific guidelines for the layout of pad cells. Remember that the primary requirement of the pad cell is to enable a reliable wire bond connection to the die.

- *Pad size:* Big enough to accept a wire bond. This limit is purely a function of the bonding machine; therefore, it is defined as part of the layout design rules.

- *Pad spacing:* Not only does the pad have to be a specific size, but pads need to be spaced far enough apart to avoid shorting two bonding wires together. This rule may be specified as pad center to pad center or from the edges of the passivation opening. Again, the design rules will specify this amount, as it is derived from the limitations of the bonding machine. This is easily accomplished by building the pad cells to abut to one another and still obey this rule. This rule effectively defines the width of the pad cell. A smaller value may be specified for double bond pads where a short is not fatal because the underlying pads are connected together.

- *Pad to other structure spacing:* In order to avoid shorting a wire bond to internal circuitry, a spacing rule is typically defined in the design rules. This rule may be a manhattan value or sometimes is specified as a region defined by a radius from the pad.
- *Pad to scribe spacing:* In order to avoid damaging the passivation or pad structure during dicing, the pad needs to be placed away from the edge of the chip.
- *No-pad zones:* Some areas may not be easily bondable, such as chip corners.
- *45-degree connection to pad:* As the pad cell is exposed to potentially higher currents from the external world, guidelines to avoid sharp corners and possibilities for concentration of charge are recommended. A simple example is to connect to bonding pads using 45-degree polygons.
- *Pad cell origin:* A trick to more easily extract pad locations for the bonding machine is to set the origin of the pad cells to the center of the large metal area defined as the bonding target. Simply by parsing the layout database for the pad cells and determining the origin of the cell, a list of bond pad locations can be generated.

Figure 5.17 shows examples of some of these rules. They are defined by the manufacturing facility, but in general, the numbers are sometimes negotiable. Design rules usually have a significant amount of tolerance built into the values. A competitive advantage may be achieved by tweaking these values, therefore using creative solutions and working with the process group to address them is usually fruitful.

Because the pad frame or I/O region is the interface to the outside world, there are special requirements to obey. Electrostatic discharge (ESD) and the issues surrounding large devices dictate most of the rules for this region.

What is ESD? *Electrostatic discharge* is the discharge of a large amount of charge into a chip. This charge can be fatal to a chip because it may physically damage transistors that are hit with the charge, much like a structure that is hit by lightning.

The magnitude of ESD can vary widely, but the duration of a pulse is usually very short. An ESD event can result in junction failure, oxide breakdown, unwanted charge injection, and fusing or opening of internal wiring.

The most common source of ESD is from the human body, when a person incorrectly handles an IC. The resulting voltage can be in excess of 20,000 V. Electrostatic damage to electronic devices can occur at any point from manufacture to field service. Damage results from handling devices in uncontrolled, low-humidity, or poorly grounded surroundings.

ESD protection structures are always built into the conduction paths from a bonding pad to internal circuitry and are part of the pad cells. These structures act as lightning rods and are intended to redirect unwanted charge away from sensitive internal circuitry.

How is this done, or what rules must be followed when implementing pad structures to maximize our ESD protection? Different techniques are used for input and output structures; these will be discussed in the following sections.

5.3.1 Output Buffers

The most complicated structure in the I/O region is the output buffer design. Here we will encounter many rules that address ESD issues.

Output buffers are large drivers that send a signal off-chip. The widths of these devices are in the range of 400 to 1,000 μm. The size will depend on the frequency, power, voltage levels, current drive and functionality, etc., of the buffer itself. These large transistors must be laid out with great attention to detail, because the area that they will require is highly sensitive and they can directly affect chip size.

It is not logistically feasible to show all of the features of output buffer design, but we will try to explain the most common problems and solutions.

Referring to Figure 5.18, we can explain a few of the output-buffer-specific rules and guidelines:

(a) Every layer, whether metal, poly, or active, is shaped as 45-degree polygons until the signal path changes layer through a contact or via. The reason is that by avoiding corners we can avoid power surges and charge concentration around 90-degree turns.

(b) The distance between the contacts and the gate on the pad side is much bigger than in the case of normal transistors. The reason is that in this way we increase the resistance through active from the contact to gate poly and reduce the voltage drop across the gate. This is done to increase the ESD protection characteristics of the transistor. In some cases, an additional implant layer is used for these regions to increase the resistance of the active layer for the same reason.

(c) The distance between the contacts and the gates on the power side could be minimal or as great as in (b), depending on the process requirements. Remember that power pins are also susceptible to ESD strikes; therefore, it may make sense to design the same ESD protection structure on the power side, too.

(d) As we can see from Figure 5.19, the current flows between source and drain in various symmetrical electric field arcs. In the case of contacts in the middle of the transistor, the field lines are equal and symmetrical.

When the contacts are placed at the end of the source/drain, the field lines converge near the edge of the active. In this region the resistance across the gate will be smaller. In case of an ESD strike, the current path will take the path of least resistance and will tend to concentrate in these areas near the edge. "Punchthrough" to the substrate from the source and drain is more likely to occur here.

End overlap gate over active

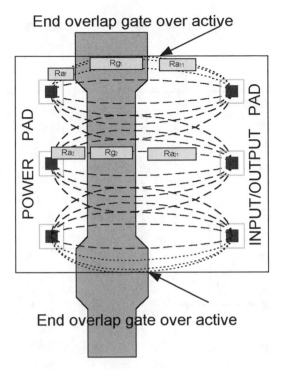

End overlap gate over active

Figure 5.18 Output buffer end overlap rule.

To avoid the concentration of the current at the edge of the transistor, we can increase the resistance of this region by enlarging the gate length. The amount we increase the gate length should be targeted so that $Ra1 + Rg1 + Ra11 = Ra2 + Rg2 + Ra21$, and the current from an ESD strike is evenly distributed along the width of the transistor.

In reality the phenomenon is much more complicated, but for layout purposes, we hope that this example of equivalent resistance is good enough.

Figure 5.19 shows two options for gate overlap termination, seen in the top or bottom of the picture. The bottom example shows the case where a poly connection to another layer is desired.

(e) Part of rule (d) includes other process-related rules specific to the edge of the active layer. Each company that manufactures chips has a few specialists in ESD, so we advise you to talk to them about any requirements they may have before completing your design.

(f) The width of the pad output connection is based on many considerations: electromigration, resistance of the metal, impedance and inductance of the package connection, and equal load between the output transistor fingers, among other things. There is a lot of approximation and "art" in choosing the right widths for the pad connection. In general, the process group typically has many "proven or recommended" values for everything in the I/O area.

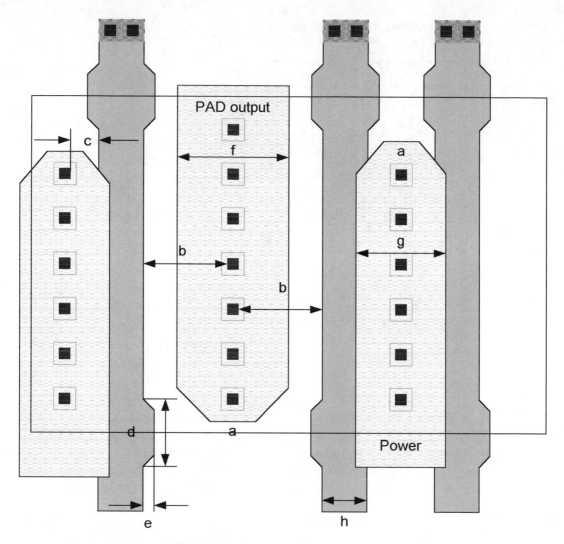

Figure 5.19   Output buffer transistor design.

(g) The width of the power connection depends in general on the power consumption of the output transistors and the issues in minimizing the power supply resistance to these transistors. In our example, all connections are shown in metal1. In today's four- to six-metal-layer processes, power to output buffers typically runs in metal3 or metal4.

A chip with 32 or 64 data buffers switching at the same time will have large chip power requirements. Note that in general output buffers are supplied with special or isolated power lines that are not connected to any other transistors and are connected directly to independent power pads. The different power supplies may be interconnected either in the lead frame or outside the package, but at the chip level, the output buffers are connected to what is called VDDQ and VSSQ.

(h)   Gate length is another important thing to remember to address. Because of the high voltage over this gate in general, the length is greater than normal. Again, process people can help us here. They test various gate lengths at various voltages and can recommend the proper gate length based on their experiments and the performance, power, and ESD requirements of the chip under design.

One last thing to note in conclusion. When working with special layout cells such as the output buffer, it is advisable to build a checklist for that region and have the cell audited in great detail.

5.3.2 Input Buffers

Input buffers accept signals from outside of the chip. In the specific case of an I/O cell, many of the ESD protection structures and techniques are built into the output transistors.

Nevertheless, input protection structures are required to protect the fragile transistors that buffer the external signal for internal use. These devices are designed and tested by the process people, so if a manufacturer is providing one to you in a specific process, the device is silicon-proven already.

In many cases the input buffer design may start before the process is well established so designing an input ESD protection device can be a tricky proposition. Figure 5.20 shows an example approach.

This solution is simply an adjustable resistor. Note that the resistor is made of an active polygon with 45-degree corners [recall rule (a) in the output buffer section]. Again, the active polygons are typically surrounded by a special ESD implant layer that has the effect of increasing the quality of the resistor.

The resistor is divided into three equal regions so that the resistance can easily be adjusted at a later date. The adjustment is used in the prototype stage only, and a fixed setting is chosen for the production design. During prototype evaluation, the resistance can be adjusted using a focused ion beam (FIB) machine that either adds or cuts the metal tracks. This has the added benefit of allowing damaged parts of the resistor to be bypassed.

There are many considerations for the design of this resistor:

- Choose the width and length of the resistor based on ESD requirements and performance characteristics.
- We highly recommend using more than one contact to connect a signal path that has the possibility of carrying high currents.
- The width of the metal line should not be minimum and must meet some kind of electromigration guideline.
- The dashed line between the metal contacts represents the main area of effective resistance. It not only demonstrates the area available that is effectively resistive, but it should be a drawn layer that can be used to create devices for LVS. LVSing these devices ensures that inadvertent changes to the cell are not made.

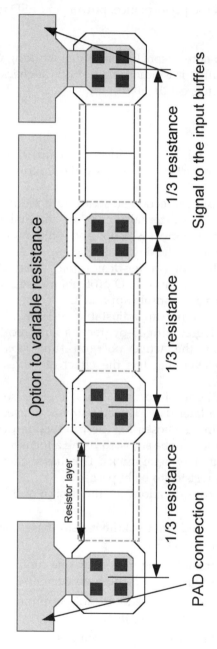

Figure 5.20 Input ESD protection resistor.

5.4   MEMORY DESIGN LEAF CELLS

Memory layout design is a real challenge to a newcomer. The design of memory-related layout cells requires detailed knowledge of both the manufacturing technology and the circuit architecture and performance issues.

In most design styles, there are portions of memory such as SRAM, but the most challenging is the dynamic random access memory, or DRAM. Figure 5.21 shows a floorplan of one implementation of a DRAM core.

First we have to see how a basic memory cell looks in a circuit (Figure 5.22). This is because the manufacturing process is most complex, as shown in Figure 5.23.

The stacked capacitor DRAM memory cell has special layers that form the memory cell capacitor. Any DRAM memory cell layout is generally very process specific, and company confidential as well.

As Figure 5.23 shows, the DRAM memory cell is very high (tall) in terms of the manufacturing process. Generally the node and plate poly layers are allowed for use only in the memory cell area. The sense amplifiers therefore use at most only two poly layers, and the topology of this area is significantly lower. "Friendly" cells are used to interface the regular patterning of the memory cells

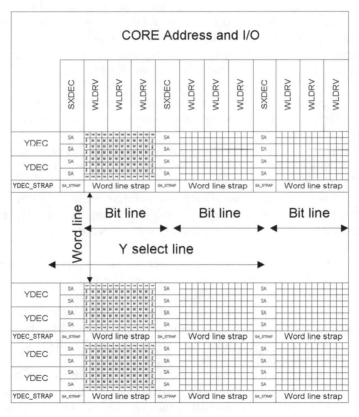

Figure 5.21   DRAM architecture.

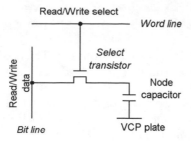

Figure 5.22 DRAM cell schematic.

to the irregular patterns outside the memory cell matrix. It is an area where the vertical topology can be gradually decreased for areas outside the array. Shielding can also be provided by connecting layers in the friendly cells to quiet signals such as power supplies.

As we explained in Chapter 4, "Layout Design Flows," the design of the memory has a very strong dependency on the layout. In fact, the layout of the memory is done first and the schematic design follows. The reason is simple: because the memory cell is repeated many, many times, it is crucial to minimize its size. From this base we build the rest of the circuitry around the matrix of memory cells. This leads us to the topic of pitch-limited layout.

Pitch-limited layout is a type of layout design where the cell under consideration is restricted in one dimension and must interface to a "leader." The leader in most cases is a repeated cell like the memory cell of a memory array. All planning effort should be focused on minimizing the size in the unrestricted dimension.

Figure 5.24 shows an example of a memory core. In this case the WL (Wordline) Driver, Wordline Strap, and Sense Amplifier cells are examples of pitch-limited layout.

In analyzing the memory array shown in Figure 5.24, note that the pitch of the cells is not the same as in the memory cell. For example, the wordline driver cell has a pitch of two cells.

The pitch matching of cells is what makes this type of layout so challenging. Whenever there is a significant mismatch in the circuitry to be implemented between two cells, inevitably the follower cells are modified in many ways to meet the leader cell's requirements.

Generally a leader cell has been highly optimized to implement a specialized circuit with minimum design rules. The result is that the dimensions of the cell are defined by a very specific number of rules.

Given these specific requirements, it is usually very unlikely that the pitch-limited cells around the leader cell can meet the same pitch. The follower cells will have different circuit requirements that will not match the critical design rules that limit the leader cell.

In the case of memory cells, it is not uncommon for the design rules to be different for the pitch-limited circuitry, since the topology is not as regular as that of the memory cells. These differences make the layout more challenging.

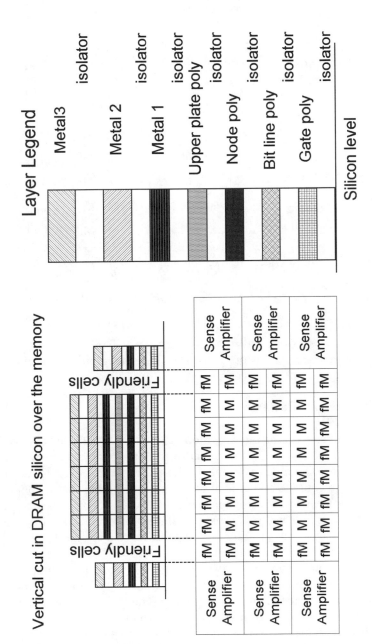

Figure 5.23 Cross-section of a typical stacked capacitor DRAM process.

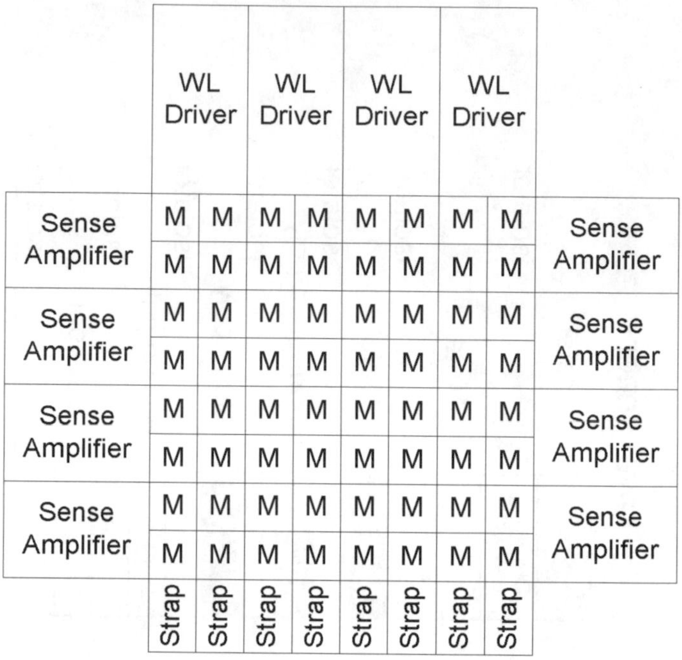

	WL Driver	WL Driver	WL Driver	WL Driver	
Sense Amplifier	M M	M M	M M	M M	Sense Amplifier
	M M	M M	M M	M M	
Sense Amplifier	M M	M M	M M	M M	Sense Amplifier
	M M	M M	M M	M M	
Sense Amplifier	M M	M M	M M	M M	Sense Amplifier
	M M	M M	M M	M M	
Sense Amplifier	M M	M M	M M	M M	Sense Amplifier
	M M	M M	M M	M M	
	Strap Strap	Strap Strap	Strap Strap	Strap Strap	

Figure 5.24 Memory array pitch-limited cells.

Subsequent sections will discuss examples of pitch-limited layout cells to illustrate the preceding concept. In all cases, it is the memory cell pitch that is limiting one dimension of these cells.

5.4.1 Wordline Strap Cells

In terms of the concept of pitch-limited layout, the wordline strap cell differs in topology from the leader cell by the addition of a contact between the poly and metal wordlines. The memory cell is built on almost an absolute minimum metal pitch without a contact, and it is at the strapping points that a contact is added. Therefore, in the wordline direction there is very little extra space to put contacts.

Why is this cell needed? The wordline strap cell is an interesting cell to understand because it is a purely layout solution to a circuit design problem, and there are many ways to implement it.

The wordline driver typically has a gate load of 1,024 memory cells. This gate load is a large capacitive load, and the resistance of the long line of gate poly makes the delay of the wordline prohibitive. The resistivity of a metal line is typically three or four orders of magnitude less than that of gate poly, so strapping the wordline regularly to reduce the wordline delay makes a lot of sense.

The frequency and placement of these straps should be determined primarily based on the polysilicon resistivity, wordline performance requirements, area, metal and contact resistance, symmetry between wordlines, and failure analysis concerns. Figure 5.25 shows different strapping schemes and the pros and cons of each.

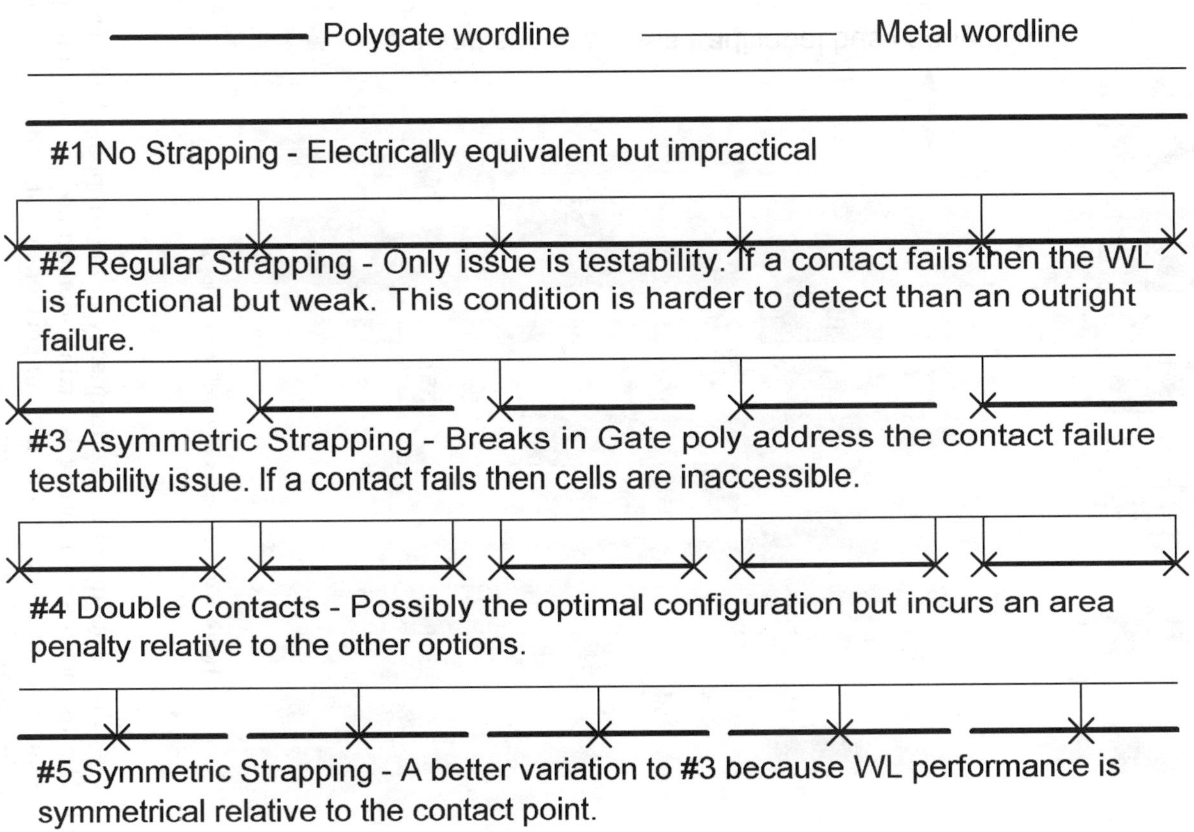

Polygate wordline ———————— Metal wordline ————

#1 No Strapping - Electrically equivalent but impractical

#2 Regular Strapping - Only issue is testability. If a contact fails then the WL is functional but weak. This condition is harder to detect than an outright failure.

#3 Asymmetric Strapping - Breaks in Gate poly address the contact failure testability issue. If a contact fails then cells are inaccessible.

#4 Double Contacts - Possibly the optimal configuration but incurs an area penalty relative to the other options.

#5 Symmetric Strapping - A better variation to #3 because WL performance is symmetrical relative to the contact point.

Figure 5.25 Wordline strapping schemes.

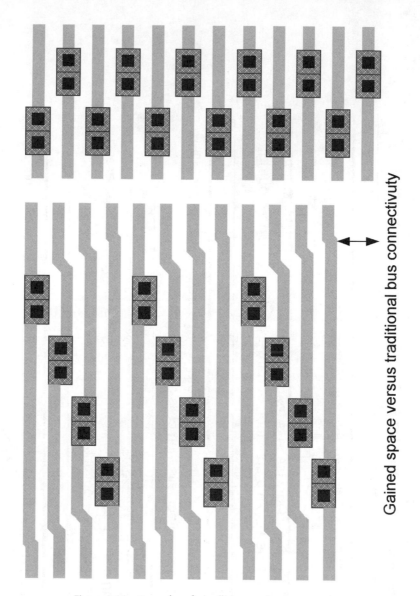

Figure 5.26   Examples of wordline strap layout.

These schemes assume that the frequency of the strapping has been determined. The frequency of strapping is mainly determined by wordline performance requirements; however, the area impact of one strapping scheme may result in alternatives being considered.

In terms of the layout of a wordline strap itself, we have to deal with the fact that across the memory cells, the wordlines run in almost minimum pitch without a contact, and the task is to put contacts in for every line.

The solution is to stagger the contact cells. By analyzing Figure 5.26, we can understand how we can gain in the width of a bus connection.

Please check the CD-ROM for a color version of the wordline strap diagrams.

5.4.2 Wordline Driver

In comparison to the wordline strap, the wordline driver is a much more complicated cell. Instead of fitting a single contact in the pitch of a wordline, we need to implement an entire driver stage! This might be a CMOS inverter or an alternative optimized specifically to minimize the area and performance requirements.

Describing the details of this process is beyond the scope of this book, but the fundamental approaches to this type of layout have been covered.

As we determined the required stagger in the wordline strap case, we generally do the same for the driver, except for transistor with interconnect staggering. Determining the minimum pitch of a transistor in a wordline driver environment has almost endless variations.

Here are some other things to consider or remember:

- Build the cells similarly to the datapath cell strategy, in that the neighbors of the cell are itself. Consider the wordline direction as the datapath direction. Use the direction perpendicular to the wordline for control lines.

- If the manufacturing process is immature or a new circuit is being implemented, most process engineers are willing to negotiate on some design rules when confronted with pitch-limited layout design. Typically, significant area penalties can be avoided, and there may be more tolerance in the design rules because the drivers are regular patterns and are close to the memory cell array where the processing is very well controlled.

- Be aware that DRAM memories have multiple power supplies such as a super voltage VPP and a negative substrate voltage VBB, and these wells and connections must be managed and planned for in addition to implementing the required circuitry.

- The staggered circuitry inherently causes asymmetry in the performance of the different wordline drivers. The goal should be to minimize these differences as well as model them in the verification of the final implementation.

As you can now imagine, the unrestricted dimension of a wordline driver can grow very quickly as we try to stagger a driver cell in the pitch of memory cells. Figure 5.27 shows a floorplan of a possible solution.

5.5 LASER FUSE CELLS

Considering the complexity of the different types of IC design: microprocessors, graphic accelerators, ASICs, etc., there are literally millions of simulations that have to be done before the chip is free of bugs (errors). The problem is that in many cases the market is pressing the design team to release the design before all the combinations of simulations are done. Another problem could be that by the time the chip is designed, the manufacturing process has evolved so that the transistor characteristics are somehow altered. Designers are constantly trying to take these issues into account. However, these precautions may not be enough to compensate for poor results in silicon.

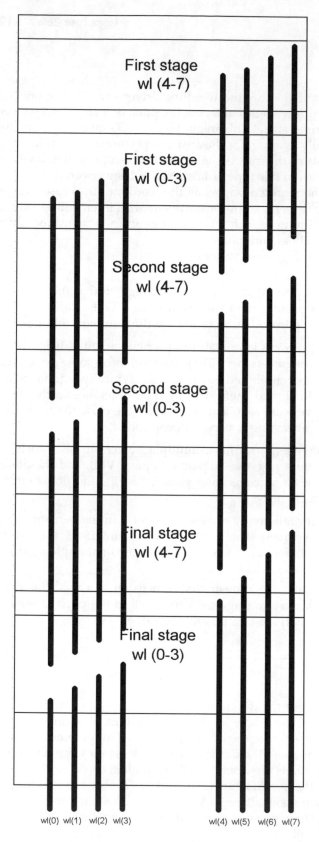

Figure 5.27 Example wordline driver floorplan.

DRAM memories are especially susceptible to process defects even though the process is highly optimized for the core layout. The memory cells, wordline drivers, sense amplifiers, and y-decoders are highly susceptible to failure for this reason. Memory designers have extensively used the concept of redundant circuitry to repair faulty circuitry, thus increasing the overall yield of the manufacturing process.

Repairing DRAMs once they have been fabricated in a production environment is typically done with laser programmable fuses. DRAM designs contain spare wordline drivers, sense amplifiers, and y-decoders that can be enabled once failures are detected and identified. A laser physically "blows" fuses that will disconnect the failing portions of the chip and replace them with the spare elements.

An alternative use of laser fuses is to provide circuit adjustment options for manufactured ICs. Similar to bond options or metal options, laser fuses can be used to configure the operation or performance of a chip.

Fuses are generally implemented in polysilicon or metal and must be built in such a way that a laser repair machine can accurately blow them out. As you can imagine, the fuses must be specially designed to isolate the impact on the rest of the internal circuitry of a laser zapping the chip. These areas need to be exposed at least temporarily, so there is a danger of contamination during this time.

As in the design of pad cells, there are physical requirements to be satisfied. For example, the fuses must be large enough for the laser repair machine to accurately program them. A list of guidelines for fuse layout would include the following:

- A design where the fuses are equally spaced is more compatible with the laser repair machine. These machines typically move to a starting point and move at a consistent speed; therefore, equal spacing of fuses is ideal.

- Similarly, minimizing the number of rows of fuses reduces the overhead of moving the laser to new starting points. Fuses will be designed in running rows placed as close as possible to each other, so movement of the laser head without blowing them should be reduced to a minimum. See Figure 5.28 for details.

- The number of fuses should be optimized so the repair time for each chip can be minimized.

- Scrambling equations have to be very clearly documented so the testing people can program the machine easily without errors.

- Special keys alignment keys for the laser repair machine are required. These ensure that the laser is exactly aligned from the start and can be done automatically. These keys usually must enclose the area where all fuses are located.

Please refer to the CD-ROM for color layouts taken from a MOSAID design. There is one example from a DRAM process and one from an ASIC process. Figure 5.29 shows another example.

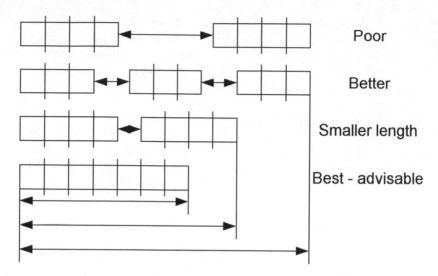

Figure 5.28 Fuse row and rules.

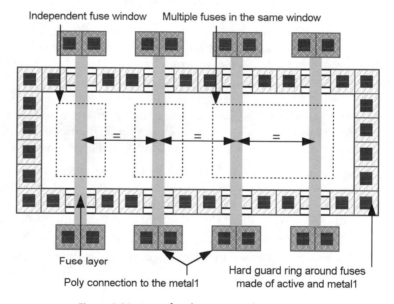

Figure 5.29 Laser fuse layout examples.

5.6 CHIP FINISHING CELLS

After all the devices related to the logical functionality of the chip are placed and verified, there is still work to be done in implementing a class of cells to finish the chip and ensure that the chip is compatible with the manufacturing process.

Examples of chip finishing cells include the following:

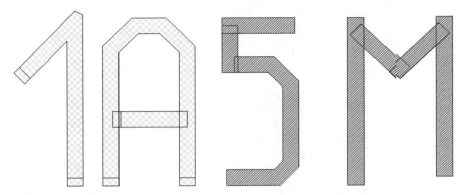

Figure 5.30 Example layer identification mark.

- Alignment keys for processing equipment
- Structures for process monitoring, dicing, and packaging
- Identification marks that may be big enough to see with the naked eye: logos, designer initials, maskright and copyright markings, process identification, layer identification (Figure 5.30)

In many cases, the manufacturer completes these tasks and the circuit designer does not have to worry about them. At minimum, the layout designer should receive all of this information ahead of time, to allocate space for the required devices; otherwise, there is a danger that there will not be enough free space on the chip to implement them.

It is important to note that after all the devices are placed and verified individually, the final full chip verification should include all of these chip finishing cells that will be included in mask making. It is easy to cause an electrical problem by placing an identification mark in the wrong place.

5.6.1 Alignment Keys

There are many alignment keys in layout design, depending on the process and manufacturing requirements. Figure 5.31 shows three examples of alignment keys.

The laser fuse alignment key is one that has to be instantiated at least three times. Depending on the manufacturer, the key is made of various layers that can be seen from the chip level. Metal and via layers are typical choices.

The NIKON keys are used to align the reticle when generating the masks on the wafer. Every layer is placed in this key; therefore, these cells will never pass layout verification because they will generate design rule errors and illegal devices. NIKON keys need to be placed in all four corners of the chip and as close to the corner as possible.

We have explained here only a few of the various keys related to fuses and packaging. There are also many keys used by process people during prototyping and robots during mass production to monitor the status of the manufacturing process.

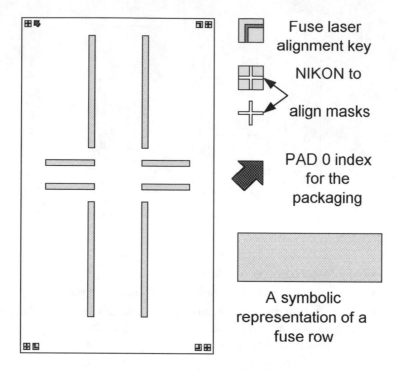

Figure 5.31   Examples of alignment keys.

There are process monitors to measure transistor performance, mask alignment, and layer resistivity and capacitance, as well as layer widths and spaces.

The placement of the process monitors is generally a function of the mask-making process in that these monitors must be a part of every mask set. If a mask set consists of only one chip, then the monitors need to be placed within the die area. In the case where multiple chips coexist within a single mask set, the monitors can be placed between die.

During manufacturing tests, bad die are identified with a black dot, and these die will be rejected when the wafer is diced.

5.6.2 Scribe and Seal Ring

Chips are never manufactured one piece at the time. They are manufactured on a large slice of silicon called a wafer. Once the wafer is manufactured, the wafer is diced into individual ICs.

Narrow channels between individual ICs are mechanically weakened by scratching them with a diamond tip. This channel is known as the "scribe" channel. The wafer is cut along the scribe with a diamond blade, or burnt with a laser. The wafer is then mechanically stressed and broken apart along the channels, thereby separating the individual ICs. Figure 5.32 documents this concept.

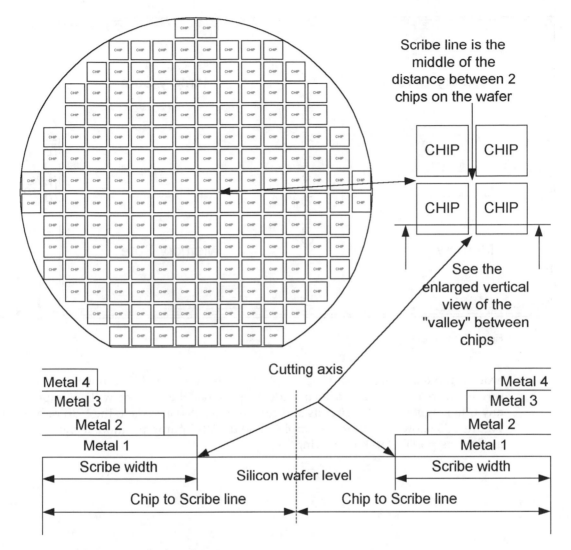

Figure 5.32   Scribe lines.

In order to protect the IC, a *seal ring* is required that is implemented around the edge of the chip.

For some chips that must have substrate connected to a source other than the normal VSS, there is another ring around the chip that is in many cases called the "seal ring." For example, many DRAMs incorporate the memory cell in isolation from the normal logic. In such cases the NMOS devices are placed in a retrograde well that is connected to a voltage source called VBB that is biased to a negative voltage. To generate this voltage and others, the memory chips have built-in charge pumps. If the external logic is connected to VBB and the entire periphery must have substrate connected to VBB, a VBB seal ring is laid out

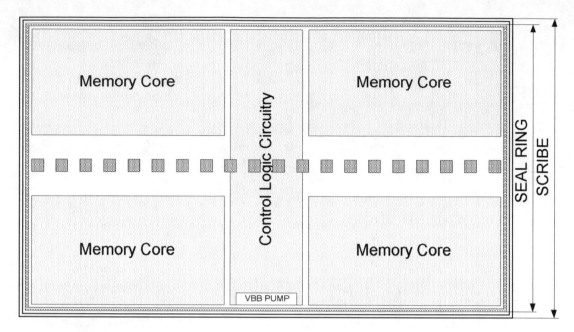

Figure 5.33   Seal ring.

around the chip. This "seal ring" is placed outside the pad area but before the scribe. The reason is that, placed outside the pad area, the seal ring will not impede any other signals and/or circuits and can be connected *directly* to the VBB pump. Figure 5.33 shows a seal ring example. Note the VBB pump position: it should be as close as possible to the seal ring.

CHAPTER SIX

Advanced Techniques for Building-Block Interconnect Layout Design

In this chapter we will apply our layout design knowledge to different and specialized types of interconnect design. Interconnect is the wires and connections between the building blocks or cells of the design.

There is an optimal order for implementing different classes of interconnect as shown in the list that follows. What is the concept behind this order? The answer is that we first implement those items that are the most difficult to insert or modify late in the design process.

Another way of looking at it is that we first implement the items that take up the most space or affect the largest areas on the die. Both answers are interrelated because the items that take the largest area to implement are those that are the most difficult to change because they are so large.

Once this is understood, the order in which we implement the interconnect portions of the design seems straightforward:

1. *Power supplies:* High connectivity and current requirements; connected to almost every transistor.

2. *Clocks:* Typically the most important and pervasive dynamic signal on a chip.

3. *Busses:* Group of signals that have to be routed together (i.e., data paths). These signals together consume a lot of routing area and are usually critical path signals.

4. *Special signals:* Non-standard-width signals or differential pairs, equal-length signals, or signals that need shielding. Implement these next to ensure that there is space to accommodate the special requirements. Special signals are discussed in Chapter 7.

5. *General routing:* The remainder of the signals. These signals are what is generally known as the interconnect signals on the chip discussed in Section 6.3.

6.1 POWER GRID

Power supply lines such as VDD and VSS are the most pervasive signals on the chip. Consider that they connect to virtually every gate and block; they each have many pins on the package; and they carry a lot of current and therefore must be sized appropriately.

The need to manage power issues very carefully in IC design has grown over time. More complex chips result in larger power grids. Voltage levels on chips have been decreasing over time from 5 V to 3.3 V, but the operating frequencies have increased. The net effect is that the power consumption of ICs today has increased. There are also many more low-power applications with cell phones, PDAs, and laptop computers. These applications really benefit from power management techniques. CAD automation for power management has been developing as well and this adds another area of expertise to study and master.

The logistics of implementing a power grid requires planning and should be one of the first things to consider when floorplanning a design. Power lines need to be planned to surround as well as flow through blocks. Adding power lines after a design has been implemented is extremely painful because typically a large and pervasive structure is needed for multiple power lines, and this is difficult to insert once a design is complete.

Remember that the goal is to provide adequate power supply connections that will meet electromigration requirements and resistance characteristics for all circuitry on the chip. Wide, short lines will meet both of these goals; however, large supply lines will inevitably consume more area. It is this trade-off that we need to manage.

In conclusion, planning and estimating the requirements for power lines should be an integral part of the layout design process.

6.1.1 Power Estimation

The first step is to calculate how much power each block will consume and therefore estimate the minimum size of the supply lines that will meet the block's requirements.

There are many ways to attack this problem, but one essential point is to do it as early as possible to avoid major rework late in the project schedule. The problem now becomes a lack of detailed information, because the designers do not know exactly how many gates will be in their block and exactly at which speeds they will be operating. Please realize that this will be an iterative process.

One very good approach is to base our estimations on a previous design, taking from it the architecture, power estimation numbers, and power routing. Extrapolating the data for the new chip we have to review:

- Differences in process constants. For example, determine if the metals are inherently more resistive or capacitive, or if they have significantly different electromigration rules.
- Check process parameters for *all* vias and contacts.

- Number of metals available, especially for the power routing. You may be able to take advantage of more or suffer with less.
- Speed of each block. Different blocks may operate at different clock frequencies and this will affect power.
- Size (in terms of number of gates) of each individual block.
- Possibility of multiple power branches, based on the noise introduced by high-speed blocks. It may be necessary to isolate noisy blocks by having separate power supply lines.
- Number and position of power pads to determine overall routability and resistance paths from the pads to the different blocks.

6.1.2 Power Supply Routing

Assuming we have been able to gather enough information from the estimation stage, it is time to plan and implement the power supply tracks to a chip floorplan.

There are many approaches to address this problem but we will describe only two basic approaches:

1. *The "root" approach:* In this case the power line starts as wide as possible, and as the power is connected to various blocks, the power supply lines become thinner, similar to the roots of a tree.
   - The width of the supply lines is based on the electromigration factor and is tapered in proportion to the quantity of current being consumed along each branch of the root.
   - This approach is used when the resistance of the supply lines is not an issue for any block along the chain.
   - Historically, most power routers routed power using this approach.
2. *The "resistance" approach:* In this case the power network may look much as it does in the root approach, but the tapering is based on a calculated resistance from the supplying pad to the specific block in the chain. The amount of resistance that is tolerable is determined by an acceptable voltage drop through the power supply line as calculated by Ohm's law.
   - The width of the supply lines is based on the resistivity of the metal used for routing the supply.
   - Number of vias is very carefully calculated to help reduce the total resistance or to ensure that via resistance is not a limiting factor.
   - The choice of metal should be based on lower resistivity values
   - Routed metal should be implemented such that changing layers   are M1 » M2 » M1 instead of M1 » M2 » M3 » M2 » M1.
   - In some specialized and fast chips, there are metal layers dedicated to power supply level only in order to reduce power supply resistance and ensure a consistent power supply level to all parts of the die.

Figure 6.1 shows the differences between these two basic styles. Most chips will use a combination of these two methodologies.

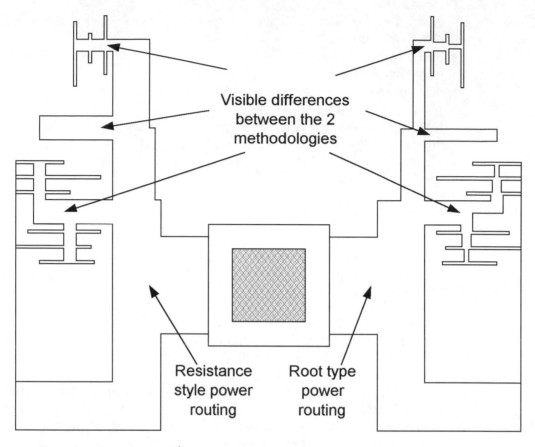

Figure 6.1 Power routing styles.

6.1.3 Strapping and Tapering

When we talk about power lines and chip layout techniques, we need to apply the concept of strapping. The idea was presented in Chapter 5 when we discussed the wordline strap cell.

The idea of a "power grid" consists of supplies routed and strapped together to form a mesh of signals. Section 6.1.2 outlined the approach of routing a power supply signal, and this section deals with the design of completing the power grid with the appropriate amount of straps and tapers. As in power routing, the amount of strapping can be determined based on electromigration or resistance, but also depends on the overall style and complexity of the design.

For example, in a standard cell block design there are specific tools that analyze power consumption and power supply resistance within a row of cells and automatically strap the power between rows to generate a solid power grid.

Certain gate array architectures have a built-in power grid with vertical strapping of power at predefined intervals.

Table 6.1 shows two examples of equivalent power grids under different conditions. Various factors can be used to determine these numbers; examples of

TABLE 6.1   Example Power Grid Table

	Strap Height in Rows	Case 1	Case 2
M1 Width Inside Cells	—	2 µm	4 µm
Maximum Strap Spacing	—	120 µm	240 µm
Vertical M2 Strap Width	4	6.0 µm	12.0 µm

those that should be used are power consumption, clock rate, average fanout, duty cycle, electromigration, and resistance.

To more fully understand Table 6.1, note the following:

- There is a direct relationship between the maximum strap spacing and the M1 width inside the cells. The relationship is defined by the resistance of the power supply line and its connection to the power supply grid. A connection to the power grid is assumed to be robust in this case.

- These numbers show an example for a 0.25 µm process and demonstrate the relationships between the different supply line widths.

Figure 6.2 graphically shows the difference in the two cases outlined in Table 6.1.

As we can see from Figure 6.2, if greater routing porosity over the cells is required, this can be achieved by paying the penalty of cell height in order to increase the required M1 power supply lines.

6.2   CLOCK SIGNALS

Typically in every design most blocks are synchronized to operate from one central global clock signal. The global clock signal usually is second only to the power supply signals in terms of its need to be routed all over the chip.

For this reason, it is important and efficient to plan for the clock signal after the power supply routing and before routing the rest. Once again, inserting a clock signal into a completed design is difficult and should be avoided.

Fundamentally, the goal of implementing a clock signal is to distribute a single signal around a large area with minimum delay. The clock signal has a large capacitive load; therefore, in order to minimize the delay, many different approaches are used.

The global clock is typically generated either directly from the pad or from an internal clock generator cell. Considerations for the layout design of the clock generator cell include the following:

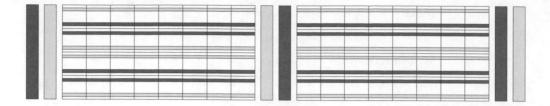

Case 1: Narrow Cell Metal, More Frequent but Narrower Straps

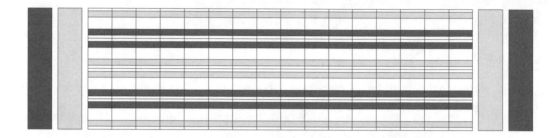

Case 2: Wider Cell Metal, Less Frequent and Wider Straps

Figure 6.2   Equivalent power strapping examples.

- *Placement:* Ideally placed near the external clock pad as well as power supply pads. The clock generator itself is a large consumer of power; therefore, it is a source of power-supply noise. This noise should be isolated from the rest of the chip by connecting the generator to independent or power pads that are nearby.
- *Buffer stage design:* As discussed in Chapter 5, the clock buffer cells can be extremely large (thousands of microns in transistor width). Each stage in the buffer chain should be designed to minimize the area and power consumption. The transistors are laid out using special methodologies for reduced power connection resistance, minimum input capacitance on the gates and most importantly minimum output capacitance.

6.2.1 Single Clock Signal

One option in implementing a global clock signal is to run a single interconnect line that originates from the clock generator.

In this case techniques similar to the ones presented in Section 6.1.2 should be used. A routing approach such as the "root" or "resistance" approach is valid.

Capacitance effects are more important in this case because the clock signal is a dynamic signal. The choice of routing layer should be made to reduce both the resistance and capacitance of the line.

Shielding of the clock line is useful to isolate other signals from the clock signal and to reduce the coupling capacitance of the clock signal.

6.2.2 Clock Tree

Another very common type of clock implementation scheme is called a clock tree. It is most common in an ASIC style of design, as the automation of generating a clock tree fits easily into the ASIC design flow.

A clock tree is a network of buffers inserted into the clock signal path in such a way that the overall delay from the generator to all destinations is minimized. Instead of one electrical signal path being optimized, the path is broken up and strategically buffered to minimize the delay. The resulting network resembles a tree in that the central clock signal branches throughout the chip using these buffers and ends up with the clock signal reaching all of the leaf cells.

Typically, the steps in implementing a clock tree are as follows:

1. An initial placement of the logic cells is completed. This ensures that the timing performance of the core logic is met.
2. The clock tree is inserted, taking into account the location of the logic cells. The buffer cells are placed or inserted in strategic places to minimize the clock delay and routing.
3. The routing is completed for all signals and optimized to meet all timing goals.

Automatic tools in an ASIC flow are available; however, the same procedure should be used in any design.

In principle, to design a clock tree a designer should consider the following:

- First define/understand the scope or extent of the clock tree. This would include items such as the total load, routing area, distance the clock has to travel, available routing layers, and routing restrictions.
- Define the constraints that the clock tree must satisfy, including minimum and maximum insertion delay and maximum skew.
- Define the way the clock tree topology will be generated, including number of levels or buffer stages in the tree and the type of buffers/inverters and fanout limits at each level. The topology can be defined manually by the designer, or automatically by a clock tree generator tool.

Figure 6.3 shows two examples of clock trees.

6.3   INTERCONNECT ROUTING

After we have solved power routing and clock tree issues, we can attack the general routing requirements. Special signal requirements will be discussed in Chapter 7. Let's review the proper order for routing signals:

1. Power supplies
2. Clock signals
3. Buses
4. Special signals—to be discussed in Chapter 7
5. General routing—the topic of this section

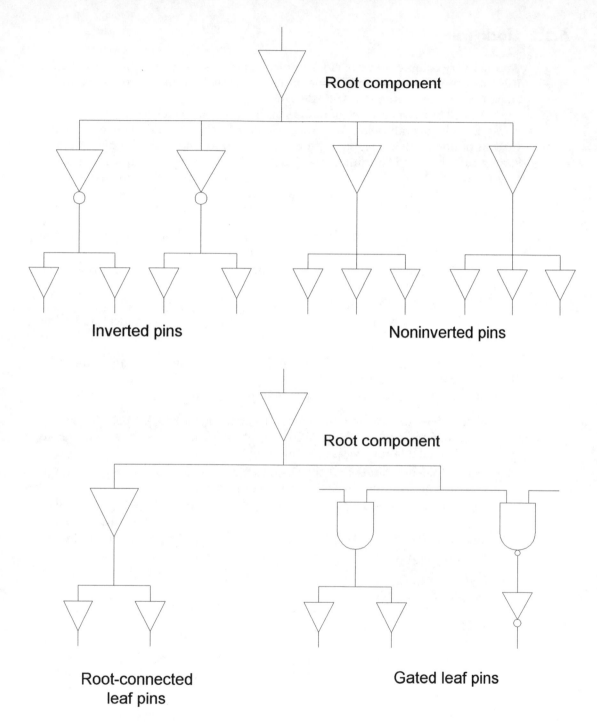

Figure 6.3 Clock tree examples.

6.3.1 Routing Plan

The goal of a routing plan is to determine the overall complexity of the routing to be implemented, identify areas on the die specifically for routing only (these areas are known as routing channels), and address potential bottlenecks or problems in achieving a completely routed design. In addition, the impact of the routing on the final chip area can be estimated. For example, dedicated routing channels can be estimated and included in a chip floorplan.

Here is a list of steps to achieve a viable routing plan.

1. Signal Estimate.   Without a final schematic or netlist, it is impossible to identify exactly the number of signals that will be required to connect all of the blocks. No matter how good the plan is, it is still a forecast of the future and will be wrong. Please understand, however, that any plan, no matter how sketchy, is infinitely better than no plan at all.

Without a finished circuit design, we can still estimate the total number of signals. This total is a good guideline for planning purposes, and if we group or relate the signals to different blocks, we can start to get a feel for areas of congestion.

A signal estimate can be based on the following:

- Asking experienced layout designers who have previously planned similar style of chips to give an estimate based on their experience. Their knowledge is invaluable in areas of congestion, blocks that have a low or high signal count by their nature, or areas that were significant problems in the past.
- A size estimate and preliminary pin list of each of the major blocks in the design.
- Signal source and destination—from where to where they have to be routed. A simple schematic diagram using blocks drawn only for the purpose of estimation (not 100 percent correct or complete) done by the architect of the chip can be very helpful. This diagram would show a preliminary location for all of the major blocks and reflect the size estimate and aspect ratio for each of them. Figure 6.4 shows an example floorplan.
- List of major busses and special signals.
- Pad list and their positions around the die.

Using all of this information, we can estimate: the location of major channels and the size of the routing channels in terms of the number of signals per channel.

If routing is allowed over the blocks, then we need to take this into consideration when defining the size and location of the channels. In this case it is useful to do a hierarchical signal plan and consider the routability of each of the blocks.

2. Establish Routing Direction.   The routing direction for each of the layers needs to be decided on a channel-by-channel basis. The floorplan is a good way to visualize the optimal choice of routing direction.

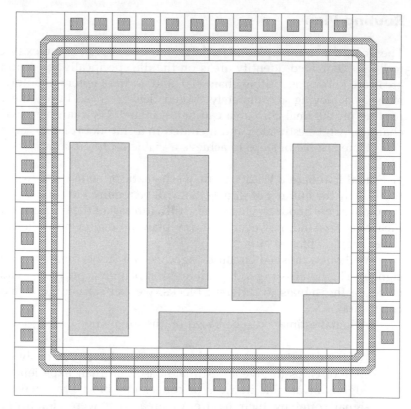

Figure 6.4 Example of chip floorplan showing signal blocks and channels.

Different scenarios for layers used in each channel should also be considered.

These issues will be explored further in the next section.

3. Contingency Plan. Finally, features and overhead to handle major changes to the design should be built into the plan. For example, spare lines and extra space should be built into the plan. The amount of overhead to deal with major changes could be determined base on the following factors:

- The novelty of the design. Newer designs will have greater uncertainty in the estimates of block sizes, signal count, and pad positions.
- The stability of the process. The routing layer design rules may change and become larger or smaller, and this would significantly affect the routing plan.
- A guideline of 10 percent is a good rule of thumb for designs of average complexity and novelty.

4. Monitor and Update. As the circuit design matures and more details on the blocks and the overall chip are available, the routing plan should be updated.

It is the process of addressing all of the foregoing concerns that results in a practical routing plan. Issues in regard to die size, congestion, and routability will be exposed and dealt with at an early stage. Once the circuit design is finalized, the routing plan will make the final implementation of the design significantly more straightforward and less error prone.

6.3.2 Channel Ordering and Routing Direction

Now that we have established routing channels in the routing plan, we can start to examine each of the individual channels in more detail. Each channel has been identified as containing a certain number of signals.

What are our concerns? Essentially, the goal of planning the channels is to prioritize and order the signals in the bus to optimize the following criteria:

- *Circuit performance requirements:* Critical path signals, signal resistance, and capacitance
- *Channel area:* Channels can be optimized to avoid jogging and unnecessary layer changes, especially as signals switch from one channel to another

In fact, if channel area is optimized, in most cases the circuit performance is optimized at the same time.

A simple procedure for manually implementing and optimizing a channel is as follows:

1. Add an unnamed path for each signal in the channel.
2. If there are critical signals that are known to be important, label them first and determine their place in the channel.
3. Label and place signals that traverse the full length of the channel.
4. Label and place signals that start or end in the channel.
5. If it is known that a signal simply goes around a row of logic to an adjacent channel, then consider adding feed-throughs to accomplish this. This is discussed in more detail later in Section 6.3.3.
6. Use the remaining space to label and place local signals. Local signals start and end in the same channel.
7. Note that the placement of cells may increase the availability of local interconnect lines as shown in Figure 6.5. Refer to Figure 6.6 for an example of placement performance reflected in channel size.
8. Leave spare placeholders for lines in the channel to anticipate new and unknown requirements.
9. Reorder the signals if necessary to optimize or minimize the number of vias or layer changes as the signals round a corner from one channel to the next. Figure 6.7 shows an ideal channel order that minimizes the overhead of signals changing channels. Not all the automatic routers have this feature!
10. Make a plot of the completed design to identify more changes that will optimize the design.

As we have mentioned, the routing layers within a channel should be determined separately for each channel. It is not necessary to define the routing layer

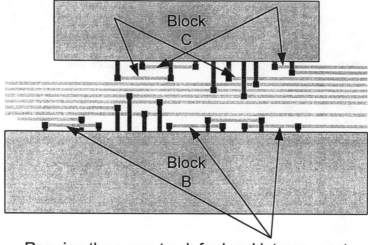

Reusing the same track for local interconnect

Figure 6.5   Local routing line sharing.

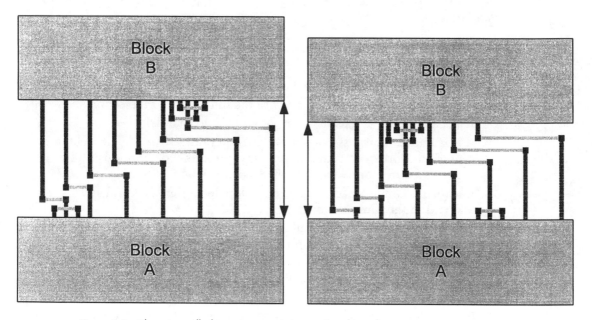

Figure 6.6   Changing cell placement to optimize routing channel.

direction for the entire chip and to enforce a set direction religiously. As we will see, channel area can be optimized by a judicious choice of routing layer for each channel, depending on the routing requirements of the signals.

Standard routing directions for each layer should be maintained for power supplies, special signals, and wide buses. These classes of signals are global and thus benefit from standardized layer assignments.

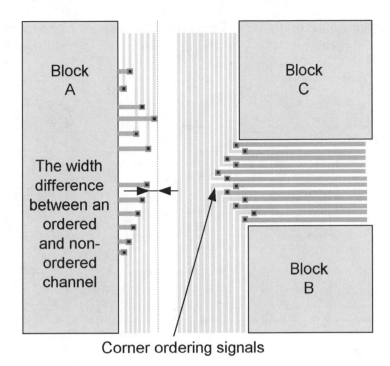

Figure 6.7 Example of channel ordering.

Key factors that determine the optimal routing layer within a channel are as follows:

- Intelligent ordering of signals for those that turn a corner as demonstrated in Figure 6.7.
- The pin layer for the cells will determine the layer that connects the signals in the channel to the cells. Note that the reverse is also important to consider, in that if a specific channel routing layer results in a smaller or more efficient layout, then the cells should be designed to take advantage of this fact.
- In general, signals to be routed parallel to a row of cells use the same layer that is routed in the same direction within the cells.
- Local signals that simply cross a channel offer a variety of choices of routing layer. Figure 6.8 illustrates three examples of this situation. We can observe that the signals in Block B are in the same order but at different locations compared to the ones in Block C. This is a typical case of only one layer in all directions. Such an approach here will provide the best size and symmetry for the bus in question.

Note the following from Figure 6.8:

- A single-layer route is only possible if the ordering of signals crossing the channel is maintained.
- Channels 2 and 3 do not suffer from the need for vias, as shown in channel 1.

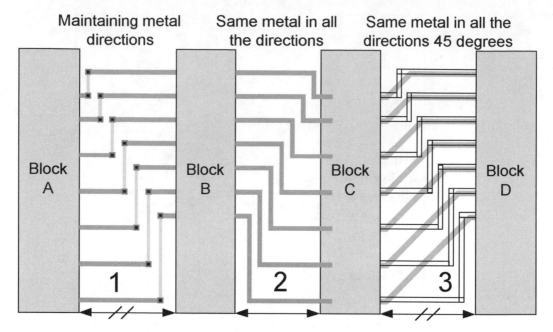

Figure 6.8 Cross-channel routing examples.

- Routing is blocked in both routing directions in channel 1. Vertical routing is free and clear in both channels 2 and 3.
- Channel 3 is implemented using only one layer, but in fact the channel is the same size as channel 1 as shown by the ghost routings. This implementation of the 45-degree signals does not result in area savings. Also, 45-degree layout is more time-consuming to generate.
- Channel 2 is the best layout of the three: the layout is symmetrical, there are no vertical routing blockages, and the channel is smaller than the others.

6.3.3 Using Feed-throughs

As the name implies, a feed-through is a routing track that simply passes through a structure without making any electrical connection within that structure. This concept was introduced in Chapter 5 as a method of passing a signal through a row of standard cells.

Let us now consider using feed-throughs in the case where we have blocks and channels. The floorplan is crucial and is an invaluable representation of the design to allow us to analyze and optimize signal routing. This optimization can occur even before we have completed the internal layout of blocks, and it is at this time that the routing analysis is most valuable. Routing requirements can be anticipated and built into the block design. The advantages of floorplanning cannot be underestimated!

Consider the scenario shown in Figure 6.9, where there are a few signals from block A to block B that are being routed around block C. There are many

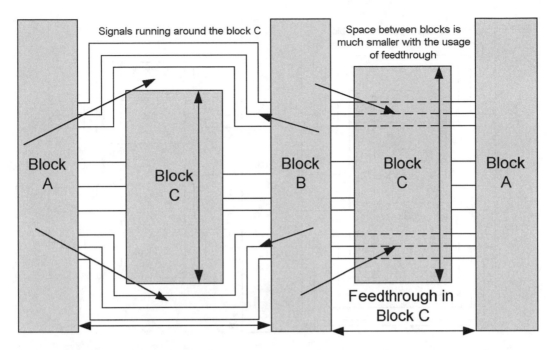

Figure 6.9 Feed-through example.

ways to resolve this problem, so let's assume that blocks A and B are complete and cannot be changed. Block C is still under development and can be altered.

The right side of Figure 6.9 shows the effect of adding feed-throughs to block C.

- Block C is longer in this example to accommodate the extra signals. This implies that there were not enough routing resources in this direction and on the desired layer to accommodate these signals.
- The signals from A to B are now much shorter.
- The distance between block A and block B is less because the channels to block C have been significantly reduced.
- Block C becomes higher because we added signal on the height dimension— or it may happen that there were enough routing resources to cope with additional signals.
- The ports in blocks A, B, and C must be aligned.
- The verification of block C will need to include the results for the additional signals. Labeling of these signals is a good idea to ensure that these will be treated as feed-throughs and will not inadvertently be used in another manner.

In an environment where many changes and additions are anticipated, there are layout design approaches that will be friendlier to feed-through after the

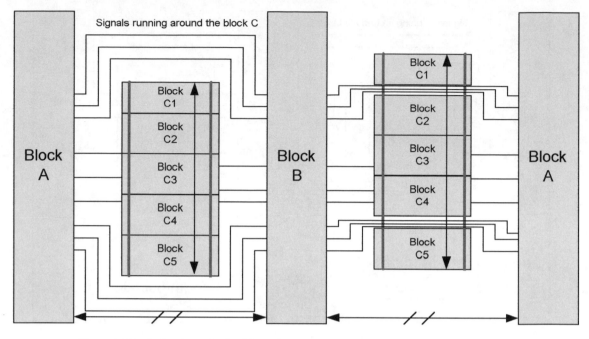

Figure 6.10 Block splitting for feed-throughs.

layout is complete. These techniques can also be used if lower level blocks are being implemented without floorplan information.

Anticipate additional routing or incorporate spare routing resources into the block design. Note the following:

- This approach may affect the block porosity and introduce overhead to the cell area. At this point it is not known if the routing space will be used!
- Identify the unused routing resources appropriately so that they will not be forgotten.
- Identify areas that are unavailable for routing to avoid mistakes.
- Ensure that the block performance is not affected by the added parasitic capacitance of the extra polygons. Similarly, if the routing resources are not used, the block performance should be verified as functioning properly.
- The simplest solution to most of these issues is to include the routing resource as part of the cell and manage the overhead of verifying the block with the extra lines. The characteristics of the block are most predictable in this case.

Another solution is to design the block in sections that can be readily split apart to accommodate extra signals. Figure 6.10 shows how this might be accomplished.

Note that by using this approach, we can add circuitry as well as routing resources.

Remember that the last two approaches are rarely required if floorplanning is used! Also, these techniques are valid for layout of small and large cells as well as blocks.

CHAPTER SEVEN

Layout Design Techniques to Address Electrical Characteristics

In this chapter we will discuss issues that have a direct impact on the electrical performance of the circuit design that is to be implemented in layout. We will cover the following:

- Resistance and resistors
- Capacitance and capacitors
- Symmetry and balancing
- More advanced layout techniques

7.1 RESISTANCE

The convention in IC design for resistance calculation is to characterize each conductor layer in terms of resistance per "square." One "square" is defined as the condition when the length of the conductor equals the width. The formula for calculating the resistance of a conductor is

$$R = \rho \times l/w$$

where ρ is the resistivity of the layer measured in $\Omega/\square$, l is the length, and w is the width of the conductor.

From this formula it should be apparent that there are two ways to minimize the resistance of a polygon (ρ is a characteristic of the manufacturing process and is not within the layout designer's control):

1. Reduce the length of the polygon
2. Increase the width of the polygon

A review of the calculation of equivalent resistance for resistors connected in series and in parallel is important to understand when good design practices reduce resistance for various layout design styles.

Figure 7.1 shows different examples of resistors connected in different ways and a calculation of the total resistance between the two nodes, A and B. It is important to note that resistors in series are accumulative and resistors in parallel reduce the effective resistance.

7.1.1 Minimizing Resistance in Transistor Design

We have already discussed resistance in terms of routing in Chapter 6. Now, let's consider the resistive effects in transistor-level layout design.

Remember that a transistor in CMOS is made of source (active), gate (polysilicon), and drain (active) regions, but to make it work we need signals connected to all three terminals. Thus, contacts to the source and drain are important to consider.

Figure 7.2 shows a fairly complex resistance model of the transistor with different resistors representing the many different current paths across the width of the transistor. Every current arc that is shown in the polygon layout is represented in the resistance model. The legend gives approximate numbers for each type of resistance and demonstrates the relative values of each of the resistors in the circuit. These numbers would be representative of a 0.25-μm process.

It is interesting to note that the active resistance is dominant and is 1,000× more resistive than metal1 and more than 10× more resistive than a metal1 contact. These numbers give us a good starting point for minimizing the overall resistance from the two metal lines: try to minimize the active resistance!

Figure 7.3 shows the effect of different contacting schemes for a transistor design to illustrate this concept.

Table 7.1 compares three cases shown in Figure 7.3.

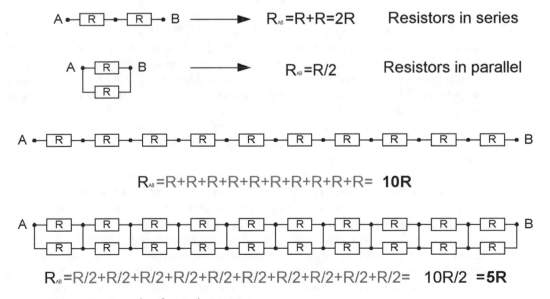

Figure 7.1 *Examples of equivalent resistance.*

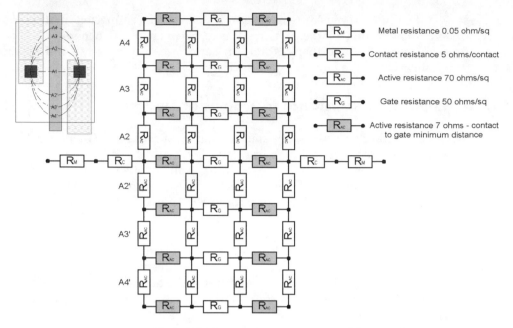

Figure 7.2 Basic transistor resistance model.

All of these examples consider the contact resistance in the analysis. Remember that the contacts are three-dimensional columns of metal or poly, and they add to the resistance of paths. The number of contacts in any connection is important to consider, because for each contact the resistance is reduced by introducing another parallel current path from the conductor. This is especially important for high current carrying signals such as clocks and power supplies.

The choice of layout style or number of contacts for transistor should be made with the application and the priority of the characteristics in mind.

For example, analog, high-speed, RF, and DRAM circuits are only a few applications that rate performance and reliability highly, so fully contacted transistors are the norm. If routability is a key issue, then using the transistor layout shown in case 2 may be appropriate. In certain places, the higher resistance exhibited in case 2 may not have an adverse effect; therefore, the benefit of routability makes this option attractive.

We would like to remind you that the calculated values should be used relative to each other, and they serve to help us evaluate the differences between these types of connections.

The transistor shown in Figure 7.4 is an extreme example that demonstrates a solution to a layout design problem where performance is sacrificed for routability. There are cases where this layout is appropriate. Certain processes will have significantly different characteristics in terms of the resistivity of the layers. For example, there are processes where the active layer is metallized and has a much lower (1/10) resisitivity than our example. In this case the performance of the transistor may be acceptable and we can take advantage of the routing channels to optimize the cell layout. These processes are more costly, but we should always be on the lookout to take advantage of special characteristics for layout design.

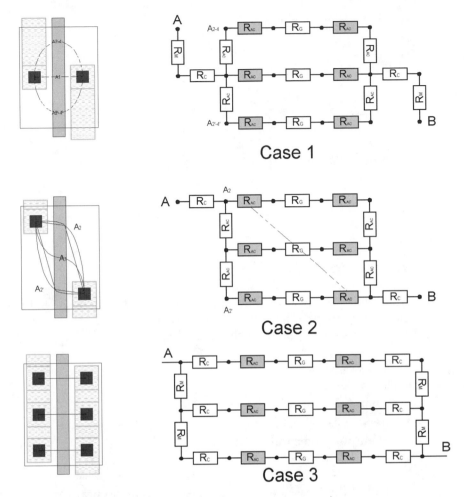

Figure 7.3 Examples of transistor contacting schemes.

TABLE 7.1 Comparison of Equivalent Transistor Resistance

Case	Total Equivalent Resistance	Comment
Case 1	56.1 Ω	Basic layout
Case 2	105.1 Ω	Higher resistance due to active layer routing to contacts Metal1 routing channel between contacts Trade-off between routing and performance
Case 3	24.7 Ω	Lowest resistance configuration No routing in Metal1 possible Highest reliability as there are multiple contacts
Case 4	450 Ω	Highest resistance configuration Four routing channels in Metal1 possible Smallest horizontal area

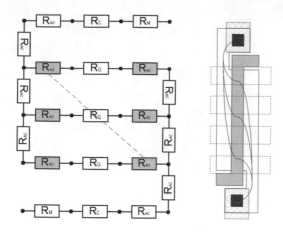

Figure 7.4   Transistor layout case 4.

7.1.2 Designing Resistors

Controlling resistance is an important concept for a layout designer to understand. In some cases an intentional resistor may be needed.

Resistors are required for the following:

- Voltage dividers
- Delay elements
- Dynamic logic loads
- SRAM cells
- ESD input protection structures
- Many analog circuit applications

In all of these cases we want a resistor and an accurate one as well.

The first step in implementing any resistor is to choose the appropriate layer. An appropriate choice of layer is possible by considering the following factors:

- Resistivity of the different layers
- Variation in resistivity under different process and environmental conditions (temperature)
- Variation in layer width under different process conditions
- Resulting area of resistor given chosen layer

In most cases gate poly is chosen as the resistor material, as its resistance is relatively high, the resistivity and width are tightly controlled, and the resulting area is not prohibitive. Some processes have a special highly resistive layer that is ideal for this application.

Resistance is calculated using squares, so to implement a specific resistance value a constant width is selected and the length of the polygon can be calculated by rearranging the formula

$$R = \rho \times l/w$$

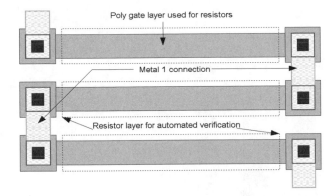

Poly gate layer used for resistors

Metal 1 connection

Resistor layer for automated verification

Figure 7.5 Typical resistors.

Wider polygons result in a longer path for a specific resistance, so the width should be chosen that produces a resistor of reasonable size. Minimum design rule polygons are usually avoided because the variation in width is most acute under this condition. It is also a good idea to standardize on a resistor polygon width for an entire chip so that all resistors will vary equally over the die. Remember to consider the effect of contact resistance!

Figure 7.5 shows examples of typical poly resistors. They are made of poly gate, connected to signals through contacts, and defined by a special resistor layer. This layer is used only for documentation and LVS purposes. It identifies the region where a resistor is recognized to establish the device for layout verification. CAD tools require that the contact layer does not overlay this resistor identification layer.

Note that resistors are prime candidates for metal options, as they are typically used in analog circuits that frequently require fine-tuning.

We can see from Figure 7.6 that these are more area efficient solutions, as they sometimes have to fit in the areas of transistors without using too much space. The disadvantage of such resistors is that the resistance is not easily calculated because of the corners in the poly layer. As a best approximation, we can use the centerline of the poly divided by the width to calculate the total resistance of the line.

In analog or RF layout designs, we may have to shield the resistor fingers from itself to avoid coupling. Figure 7.7 shows an example of such a resistor.

In some very sensitive circuits when there are two resistors connected to signals that are switching on opposite clock edges, we can balance the coupling effects between the two resistors to ensure that they both operate in a similar manner. These are referred to as balanced interlaced resistors. An example is shown in Figure 7.8.

7.2   CAPACITANCE

The definition and sources of capacitance are important concepts for a layout designer to understand. In some special cases the circuit schematic requires capacitance, but in general the emphasis in optimizing a layout design is to minimize the parasitic capacitance inherent to the different layout structures.

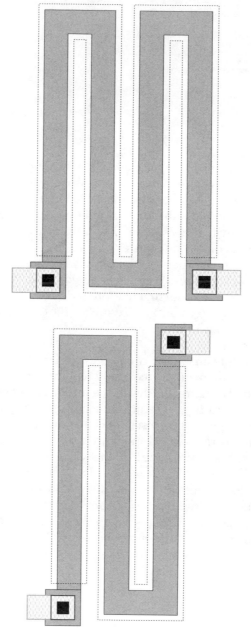

Figure 7.6 Efficient resistor styles.

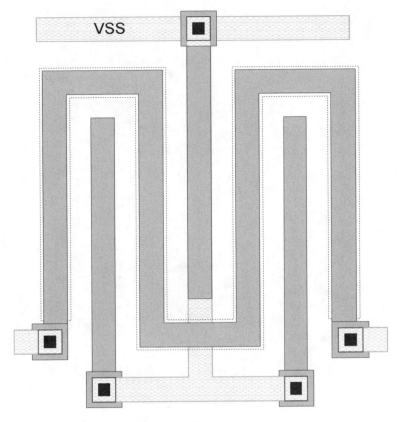

Figure 7.7 Shielded resistor.

Capacitance affects several different characteristics of a design. When two equivalent designs are compared, the design with the higher capacitance will have a resulting increase in all of the following:

- Signal delay
- Power consumption
- Coupling effects to and from neighboring structures

A review of the definition of capacitance will give us an understanding which good design practices reduce capacitance for various layout design styles.

The general formula for the calculation of the capacitance of a conductor is

$$C = \varepsilon \times A/d$$

where A is the surface area of the specific conductor, d is the physical distance between the conductor and the reference node, and ε is a constant representing the characteristics of the insulating layer between the conductor and the reference node. Figure 7.9 shows the theoretical definition of a capacitor.

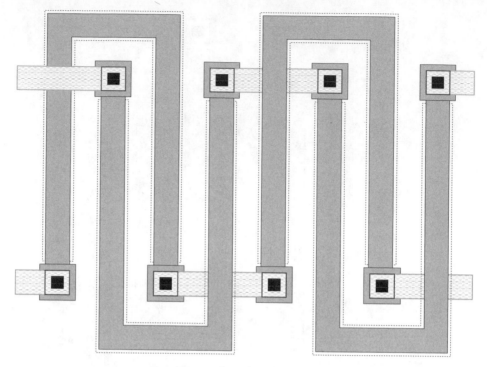

Figure 7.8 Balanced interlaced resistor.

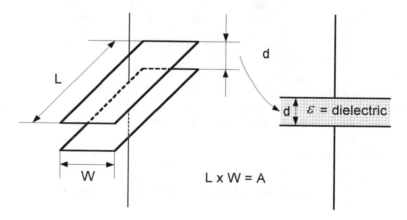

Figure 7.9 Capacitor definition.

From this formula it should be apparent that there are two ways to minimize the capacitance of a signal (ε is a characteristic of the manufacturing process and is not within the layout designer's control):

1. Reduce the area of the capacitor—this means reducing the overlapping regions of the two "plates" or polygons
2. Increase the distance between the plates of the capacitor

The effect of a capacitance C on a signal is to limit the rate of change in voltage (dV/dt) on the line by requiring more charge or current I according to the formula

$$dV/dt = I/C$$

It is from this relationship that we derive the delay formula $t_{delay} = RC$.

7.2.1 Designing Capacitors

As we mentioned, there are certain cases where a capacitor is an integral part of the circuit design. Examples include the following:

- DRAM memory cell
- Power supply decoupling capacitors
- Power supply generator reservoir capacitors
- Delay chains
- Specialized analog circuits such as switched capacitor applications

How can capacitors be predictably designed?

As described in Chapter 5, a DRAM memory cell capacitor is simply the overlap of two layers that has been optimized to be very close together. In this case the d distance term has been minimized to increase the capacitance value.

Within a standard CMOS process, the choice of layers to implement an effective capacitor is really limited to one case. By design, a transistor is manufactured to have a very small distance d between the gate poly and active layers. The source and drain nodes connect to one terminal of the capacitor and the gate node is the other.

Figure 7.10 illustrates two implementations of an NMOS transistor-based capacitor. One is in the substrate and the other has been drawn within an N-well. The N-well transistor results in a higher effective capacitance because of the lower threshold upon which the transistor starts to operate.

Power supply decoupling capacitors may be the most common use of intentional capacitors, so a few comments about them are warranted. These capacitors are connected between two power supply nodes (such as VDD and VSS) to "decouple" the two nodes and provide dynamic charge to what can be very noisy signals. This decoupling serves to stabilize the power supply voltage and increases the reliability of the chip operation.

In order to provide a measurable amount of charge, these capacitors need to be pervasive on the chip, and the effective size of the total capacitance can be in the nF or nanoFarad range. Some planning is required to achieve an effective implementation; however, power supply lines are generally readily available in many places, so it is not too difficult to find space once the regular circuitry has been implemented. It is also recommended to isolate these large transistors with guard rings to avoid noise coupling to unwanted circuitry.

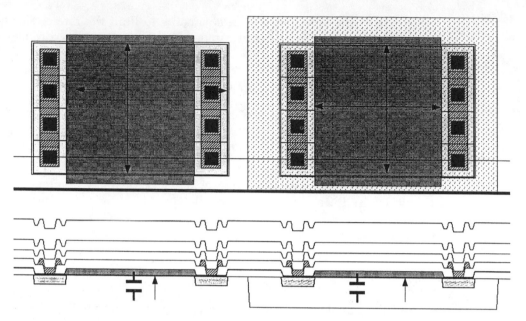

Figure 7.10 Transistor capacitors.

7.2.2 Minimizing Parasitic Transistor Capacitance

There are many parasitic capacitances inherent in a transistor, as shown in Figure 7.11.

Figure 7.12 shows a few examples of transistors designs that significantly reduce the capacitance of the drain C_{DB}. Comments for the figure are given in Table 7.2.

In general, if minimum design rules are used, then the transistor capacitance can be optimized using the techniques shown in Figure 7.12.

7.2.3 Interconnect Capacitance

Parasitic interconnect capacitance is impossible to avoid, and in most cases it is an effect that must be dealt with by compensating for the load by proper circuit design.

It is mainly for the modeling and calculation of the interconnect capacitance for every node in a design that extraction tools are used. This section will present the concepts behind the sources of capacitance that the extraction tools try to model and calculate.

Consider that a chip is manufactured from many layers placed on top of one another. The bodies near a particular polygon are numerous, and each pair of near bodies creates a parasitic capacitor. These days it is recognized that 70 to 80 percent of the total capacitance of any particular node is due to the parasitic capacitance of the interconnect routing. The shift toward dealing with interconnect loading rather than transistor loading has come about as a result of increased die sizes, increased number of interconnect layers, and smaller line pitches.

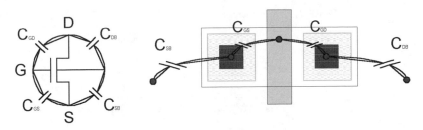

Figure 7.11 Transistor capacitance model.

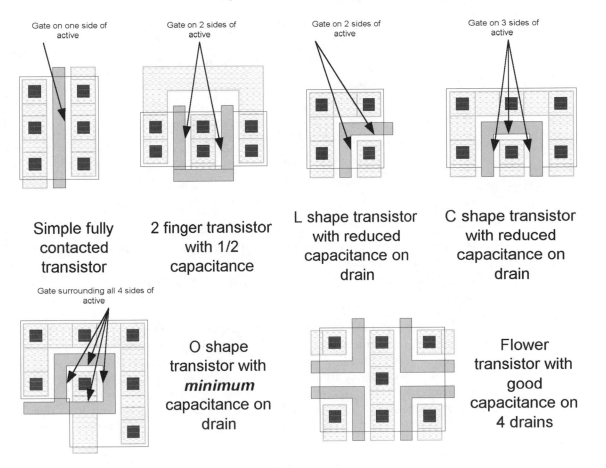

Figure 7.12 Minimum capacitance transistors.

Once an understanding of the source of interconnect capacitance is reached, then it is a matter of addressing this issue using advanced techniques. Figure 7.13 gives a three-dimensional view of various scenarios to illustrate the different sources of interconnect capacitance.

One presentation on the CD-ROM shows an example of extraction that is trying to bring the real data back to the circuit design.

The capacitance of any signal is fundamentally composed of three types of elements, as shown in the Single Line Capacitance Model portion of Figure 7.13:

TABLE 7.2 Comments for Figure 7.12

Example	Comments
Simple	• Standard layout
Two-finger	• C_{DB} at least halved when compared to standard layout
	• Maximize use of even-fingered transistors
L-shape	• Small drain, good for devices around the size of a single contact
C-shape	• Small drain and ability to easily connect both ends of the gate
	• Can be used in datapath or bus interface applications
	• Absolute minimum drain capacitance good for datapath applications
O-shape	• Basis for "chocolate" or "waffle" transistors; refer to CD-ROM for full example
	• For very large transistors, electromigration rules will limit its use
Flower	• Extension of L-shape for four output nodes

1. *Parallel-plate capacitance*: This is the simple capacitance model that was described in the beginning of Section 7.2.
2. *Fringe capacitance*: This capacitance is caused by the electric field induced as current flows down the line.

The modeling and calculation of this capacitance is well beyond the scope of this book but it is important to know that the fringe capacitance can be a very high proportion (~50 percent) of the total interconnect capacitance!

Suffice it to say that the fringe capacitance is dependent on the distance d of the signal from the body in question.

It is only recently that extraction tools have been developed to address this issue, as they historically have been limited to parallel plate capacitance type of models.

3. *Coupling capacitance*: Coupling capacitance is defined as a capacitance from one signal node to another.

A simplistic model to visualize the coupling capacitance would be to use the parallel-plate and fringe capacitance calculation between the signals in question.

The Coupling Capacitances portion of Figure 7.13 tries to illustrate this point. Near-body capacitance can be to other signals or to a variety of structures such as power supply nodes and the substrate, as shown in the figure.

Accurate modeling of coupling capacitance is limited to very specific applications simply because the tools and methodologies behind simulating and

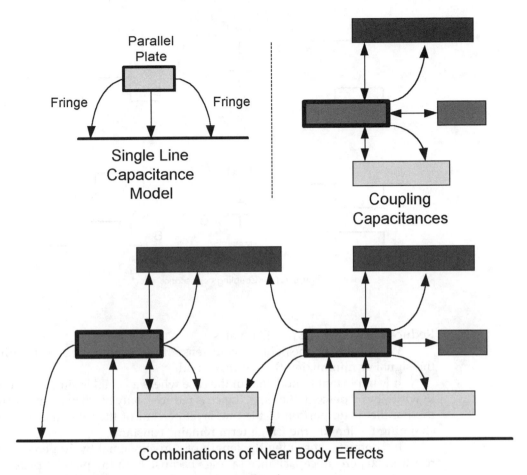

Figure 7.13   Interconnect routing capacitance examples.

extracting the size of networks that would take into account all near-body capacitance are not practical.

It is simplistic to assume that the capacitance of a node is always relative to ground. Considering coupling capacitance is crucial in the scenario where the two signals on either side of the capacitor are changing voltage in the opposite direction at the same time. This scenario is shown in Figure 7.14.

In this case the effect of the coupling capacitance to the delay of the signal is *double* that of the case where the reference node is a static signal. The driver of the line is trying to drive the signal one way and is fighting the parasitic capacitance of the line. As the reference node voltage of the capacitor is changing in the opposite direction, it couples into the line and the effect is to delay the signal even more.

Conversely, if two signals are being driven in the same direction, then they help each other.

Now that we understand the sources of parasitic capacitance, what are some techniques to reduce the different capacitive loads on a given signal?

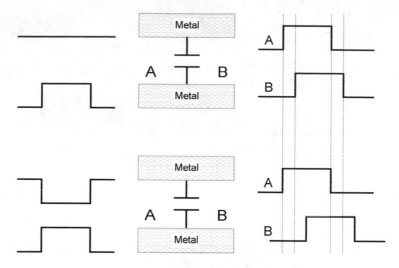

Figure 7.14 Coupling capacitance effect.

- Reduce the area of the parallel plates.

   What does this mean in a layout sense? Either shortening the length of the signal or minimizing the width, or both.

   It is important to note that in the case where a signal is simply reduced in width by 50 percent, the capacitance is reduced only by 25 percent because the parallel plate portion of the line is 50 percent of the total and, without changing the length, the fringe term remains constant.

   In contrast, shortening the length of a given signal by 50 percent will result in 50 percent capacitance savings because both the parallel plate and fringe terms are affected equally.

- Reduce the distance d or dielectric distance between the parallel plates.

   Again, what does this mean in a layout sense? Whenever possible, route critical signal lines in empty channels and minimize the amount of area that the signal is routed over or under other layers.

   As an example, assuming an empty routing channel was used, using the topmost routing layer would have the least capacitance relative to the substrate terminal when compared to other layers.

   An implementation using the top routing layer and appearing as close to the Single Line Capacitance model shown in Figure 7.13 is the ideal to shoot for.

- Increase the spacing between signals on the same layer to address the coupling capacitance between them. We call this signal spreading, and it is exceptionally useful in areas where the routing congestion is low. Please review the Sagantec presentation on the CD-ROM.

- In the case of differential signals, implement a "twisting" scheme that reduces the coupling effects of the adjacent lines by ensuring that any coupling affects both signals of the pair equally. Figure 7.15 demonstrates this concept for signal XX and introduces the concept of shielding.

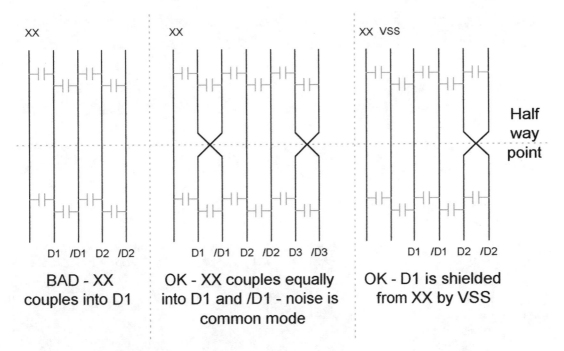

Figure 7.15 Differential pair twisting to reduce signal coupling.

- Shield critical signals with a signal that remains at a static level. This technique eliminates the possibility of the worse-case coupling scenario illustrated in Figure 7.14.

The concept of shielding signals was imported from the printed circuit board (PCB), where some signals that are supposed to provide the circuits with a reference voltage are isolated from interference to a greater extent than other signals. One example of shielding is shown in Figure 7.15, but there are more elaborate techniques. A signal can be shielded on both sides, as shown in Figure 7.16. In this case we are isolating the signal from influences on the same layer.

Figures 7.17 and 7.18 show examples of shielding of greater sophistication in that the signal in question is completely surrounded, including in directions above and below the conductor. The capability of shielding is dependent on the available routing, as shown by the two figures.

7.3   SYMMETRY

Predictability in the behavior or performance of a design can be thought of in many ways. For example, in measuring the timing characteristics of a circuit, it is desired to meet an absolute performance target. Often it is desired that two layout designs be implemented identically so that the performance characteristics of the two circuits relative to each other match.

Following are examples when symmetry is routinely used:

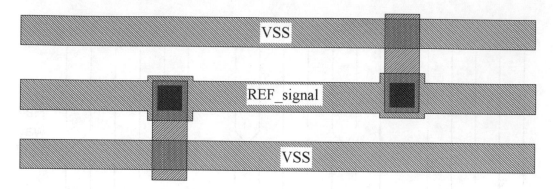

Figure 7.16 Simple shielding example.

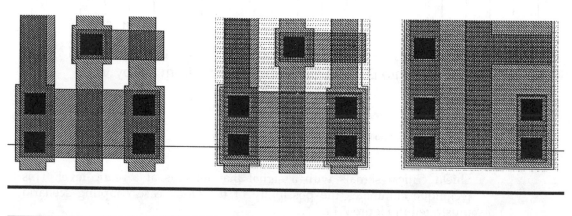

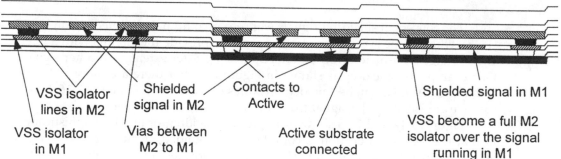

VSS isolator lines in M2

Shielded signal in M2

Contacts to Active

Shielded signal in M1

VSS isolator in M1

Vias between M2 to M1

Active substrate connected

VSS become a full M2 isolator over the signal running in M1

Figure 7.17 Shielding options in two-metal process.

- Differential amplifiers require operational matching between halves of the cell layout
- Datapath and memory array circuits require identical layout for each row and column
- In layout designs where parallel structures are used to build up a cell, such as a multiple-finger NAND gate, tweaking any series connectivity in the different parallel paths will remove asymmetries between the series elements

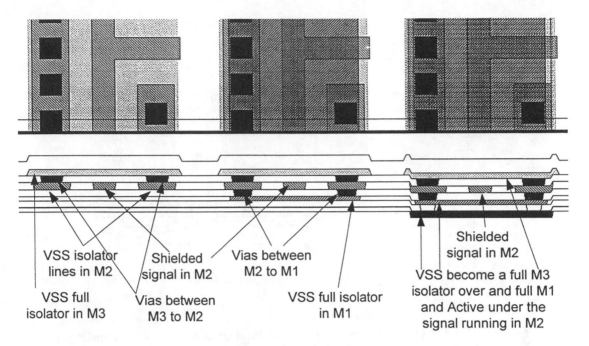

Figure 7.18 Shielding options in three-metal process.

- Differential pair routing requires matching characteristics between signals
- Designing signal and clock paths to meet flip-flop setup and hold times is an exercise in matching signal paths to the clock path

The best way to ensure that two circuits behave identically is to use the same layout cell in both cases. Signal symmetry is achieved by designing two signals to have the same length, width, load, and coupling environment. These concepts are discussed in more detail. Note that in extreme cases all the techniques described in this section can be combined for very sensitive applications.

7.3.1 Symmetrical Layout

In many analog, RF, or sensitive digital designs, two halves of a particular design are electrically equivalent, and it is desired that each half perform identically. A differential amplifier is an example where the circuit's function is to distinguish signal differences between two input signals.

A very simple technique to achieve almost perfectly symmetrical layout is to use one cell twice. The minor differences between the two cells should be implemented on top of the cell. Figure 7.19 illustrates this approach.

Advantages and disadvantages of this approach are as follows:

- Symmetry is guaranteed and easy to implement
- The benefits of cell-based layout come into play. Changes to both halves are done in one cell.

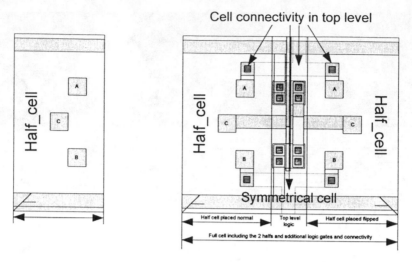

Figure 7.19 Symmetrical layout example.

- The cell planning is a little more complicated because the symmetry point has to be understood and the layout implemented with this in mind.
- There may be some area or routing overhead because of the symmetry, but polygons along the line of symmetry are usually shared.

A practical tip: Use the "Edit in Place" mode built into almost all layout editors to develop the base cell. Edit in place one of the two base cells from the top level cell so that changes made will appear in the other half instantly.

7.3.2 Balanced Layout

We define a balanced layout as a design that has symmetrical performance as a result of intelligent connectivity within its structure as well as a symmetrical layout.

Often, implementing balanced circuitry has the following effects:

- Reduced power in a world where designers are dealing with the large power consumption of today's chips and are trying to avoid needing fans to cool chips down to the optimum working temperature.
- Timing symmetry. This is especially important in analog and RF designs, where the timing of each switching device is critical.
- Definition of more detailed connectivity models to accurately capture the balanced nature of the design.

Analog design technique is the father of balanced layout, where a very old method called "balanced devices" is applied to IC design. The example shown in Figure 7.20 illustrates the concept for a two-input NAND gate, and the result is that the delays between both inputs to the output are the same.

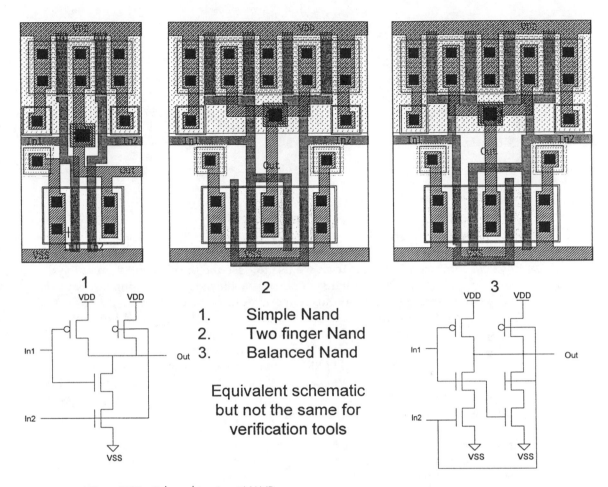

Figure 7.20   Balanced two-input NAND.

The schematic shown on the right side of Figure 7.20 shows exactly how the schematic is defined for Example 3. Many LVS layout verification tools have algorithms to recognize NAND gates within the layout. The layout NAND shown in Example 3 is not often recognized as a NAND and creates discrepancies when compared to a regular schematic NAND. The reason is that the order of the series connections within the NAND is reversed. Functionally, they are equivalent and in fact balanced. In this case the schematic must be altered to reflect the correct connectivity in order for the LVS to pass.

Balancing circuits is not always as straightforward. Balancing series devices is more difficult when dealing with more than two transistors connected in series. Figure 7.21 shows an example of three series gates to illustrate this concept further.

In order to balance the series connections, each input is connected to a transistor in each of the three positions: close to out, center, and close to power. This is only possible if there are three parallel series chains; therefore, introducing balancing to a layout may incur significant overhead.

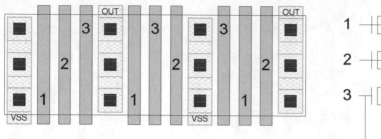 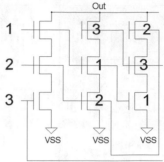

Figure 7.21 Balanced NMOS series transistors.

7.3.3 Physical Compensation

Physical compensation is the term we use to describe the concept of layout symmetry applied to signals. Signal symmetry is achieved by designing two signals to have the same length, width, load, and coupling environment.

Designing signal and clock paths to meet flip-flop setup and hold times is an exercise in matching signal paths to the clock path. In some cases where the timing margins are extremely small, physical compensation techniques are used.

Synchronous DRAMs (SDRAM) are an application where input setup and hold times are required to be very well defined. Circuit performance needs to be guaranteed under all voltage, temperature, and process conditions. Physical compensation is appropriate in this case.

As we have mentioned, signal symmetry can be achieved by mirroring many structural features of the signals among the group to be compensated. This should include numbers of vias; matching of routing layer(s) and the length of the sections in each layer; and shielding each signal equally.

A short list of steps to implement physical compensation among a group of signals might be as follows:

1. Route signals as you would any other signal, and reserve space for length compensation and shielding lines. Route all lines using a single width of line.
2. Determine the longest line among the group to be compensated and increase the length of all other signals to match. Serpentining signal lines is appropriate as long as adequate shielding is maintained.
3. If different routing layers are used, match the length of interconnect on each layer for each signal. It is not necessary to place the different layer routing in the same place along the line, but it is important to use the same number of vias within each signal.

   Match the relative transistor loading or fanout for each line. This means that the ratio of the size of the transistor load relative to the driver should be the same for every line. Additional transistor loads should be added to the appropriate signal to match the fanout ratio.
4. Table 7.3 illustrates the concept of compensation load calculations.
5. Run layout extraction tools to verify the results and adjust the layout if necessary.

TABLE 7.3 Example Compensation Load Calculation

Signal	Driver Size	Load	Fanout Ratio	Compensation Load
A	100/50	500/250	5	—
B	20/10	40/20	2	60/30

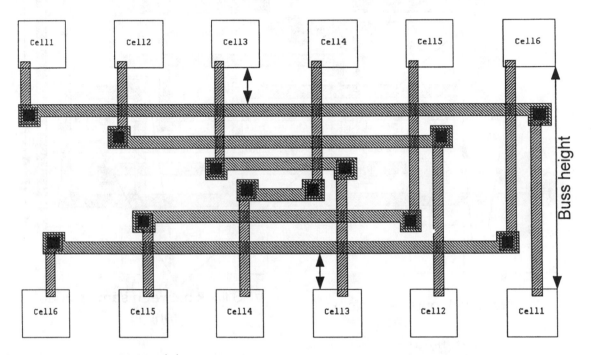

Figure 7.22 Simple bus routing.

Figure 7.22 shows a simple routed group of signals before compensation is applied. Note the arrows. The small arrows indicate areas of overhead space allocated for the physical compensation that will be applied later. The long arrow at right shows the size of the bus.

Figure 7.23 shows the same layout after compensation. Note that the final height of the bus is greater than before, indicating that the plan was too aggressive (the small arrow indicates the difference).

7.4 SPECIAL ELECTRICAL REQUIREMENTS

7.4.1 45-Degree Layout

We have already explained that polygons or paths drawn with 45-degree angles use more database space to store the polygon information. For this reason layout designers should try to avoid drawing polygons at an angle other than 90 degrees. However, there are many occasions where 45-degree layout can result in a smaller or more reliable design.

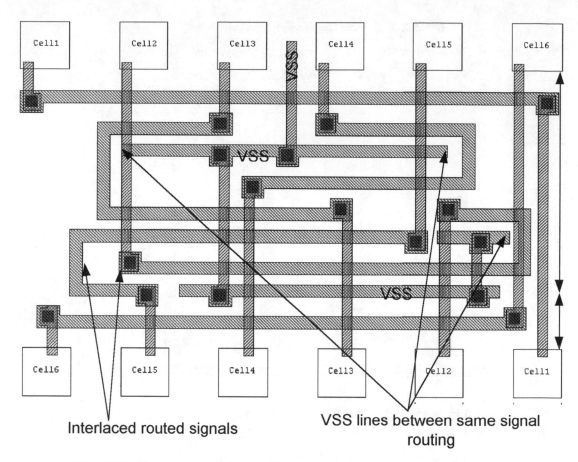

Figure 7.23 Physical compensation example.

The electrical characteristics of 45-degree polygons also make them very useful in certain areas. When large currents are being drawn through the conductor, 90-degree corners are stressed. Using 45-degree polygons alleviates this problem.

There are a few places where 45-degree layout is highly recommended or obligatory:

- Very wide power lines that are routed around corners. A guideline for very wide is more than 30 to 40 µm. In this case the stressing of the metal by high currents is reduced.
- Wide lines or signal buses turning in opposite corners. See Figure 7.24 for details. Significant area savings can be achieved.
- For signals routed close to the corners of the chip to avoid stress induced as the chip is diced.
- I/O cells to reduce power surge in the corners and possible spikes in case of ESD as discussed in a previous chapter.
- To save space when designing very repetitive structures or pitch-limited layout.

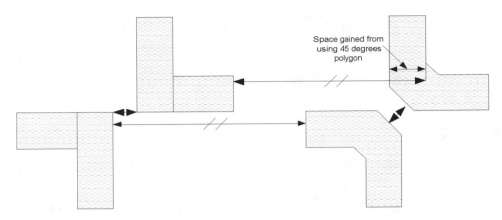

Figure 7.24 Optimal routing of wide signals.

- Optimizing transistor layout. Most of the reasons for 45-degree layout are easy to understand, but how we can gain in the design of a transistor is not such a simple concept, so further explanation is warranted.

Bent transistors have already been shown in Section 7.2.2 in the context of minimizing drain capacitance. These transistors are not optimal, because the 90-degree corner regions of the transistor are ineffective as transistors in that they generally do not pass a lot of current and can be considered to be a dead part of the gate. The corners of a bent transistor should ideally be discounted as useful transistor width.

In fact, layout extraction tools are not accurate in this area, as they use a simple algorithm that employs the center line of the gate as a measurement of gate width. The effective width of a bent gate should be reduced in size by the number of corners times the gate length.

Figure 7.25 shows the difference between 45-degree transistors and 90-degree transistors. A gate width increase is achieved.

The 45-degree gates are used extensively where any small gains in cell size make an impact on chip size. For example, bent gates are used in the pitch-limited cells that are repeated many times in a chip. Figure 7.26 shows the example of a wordline driver transistor that is very similar to many I/O cells in a PLA or compact register file. Using 45-degree polygons results in minimum size design in both X and Y directions, but changes to such transistors are not fast or easy.

7.4.2 Electromigration

Electromigration is defined as the molecular displacement of atoms caused by the flow of electrons over extended periods of time. What this really means is that metal lines will eventually break and create an open circuit by this effect.

Poorly designed parts will fail in the field after a period of flawless operation because of a high degree of susceptibility to electromigration. With decreasing scales of integrated circuits, concern has grown over the susceptibility of metal lines, operating with high current densities and elevated temperature, to degradation due to electromigration.

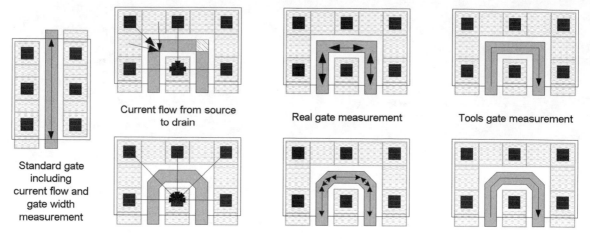

Standard gate including current flow and gate width measurement

Current flow from source to drain

Real gate measurement

Tools gate measurement

Figure 7.25 Simple examples of bent gates.

Electromigration is most prevalent in aluminum and aluminum-alloy interconnect layers. This is one reason why new metals are being introduced into the manufacturing process that will have no electromigration limitations (i.e., tungsten, copper, etc.) and very low resistance.

Electromigration failures are avoided simply by increasing the width of current-carrying metal lines so that an open circuit is avoided. As electromigration is dependent on the amount of current flowing within a conductor, a set of guidelines is used to guide layout designers to address this issue effectively without incurring significant area overhead.

There are three basic scenarios to be considered for the application of electromigration guidelines: (a) DC current, (b) AC current where the electrons nominally flow in one direction only, and (c) true AC current (electron flow is bidirectional).

In the ideal scenario, the engineer designing a circuit evaluates electromigration susceptibility for every signal. The schematic is annotated for the layout designer indicating the signals that require special attention. However, this method is not very practical, as it is work intensive and is prone to oversights and omissions. In an attempt to simplify compliance with electromigration guidelines, most companies prefer to generate a simple table for the layout designers' reference.

Before we start using such a table, it is important to understand what is a unidirectional and a bidirectional current as it pertains to a circuit design. This knowledge is necessary to let us implement the layout appropriately based on reading a schematic.

Figure 7.27 shows a inverter schematic and a corresponding layout that are annotated with the flow of current. The areas of unidirectional and bidirectional currents are also indicated.

Please take the time to understand the direction of the current flow in the PMOS and NMOS transistors. As we can see, the assumption is that the output connection will be made to the hashed polygon or to the bolded rectangle on the

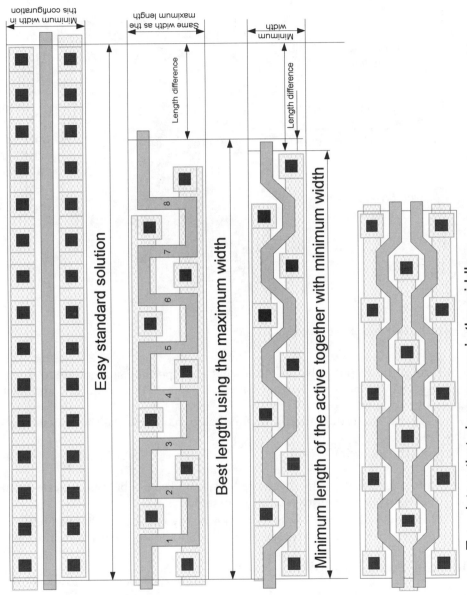

Transistors that share power in the middle.

Figure 7.26 Bent transistors to minimize area.

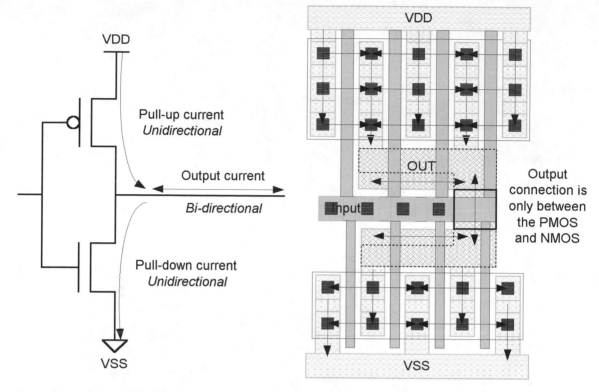

Figure 7.27 Electromigration currents.

right side of the output polygon. Areas that are especially susceptible to electro-migration would be very larger drivers such as output buffers and clock generators or buffers.

A typical electromigration guideline table may look like the one shown in Table 7.4.

Note the following:

- The driver size is an effective inverter size and so can be used for other gates if an equivalent inverter size is calculated.
- Electromigration rules apply to power supply nodes as well! This is usually not a problem, since power supplies are wide and the connections are robust.

7.4.3 Multiple Power Supplies

Chip designers are trying today to reduce power consumption of their chips, mostly because they want to put them in portable devices. As the gate size goes down, the power supply voltage is also reduced. Today there are many standard power supply voltages, but for 0.25-μm processes the power supply voltage is in the range of 3.3 to 2.5 V.

The power supply voltage level on the board or on the pins of the package is not necessarily the voltage that the chip will work on internally. As an example,

TABLE 7.4 Example Electromigration Table

Driver Size	Iavg	Bidirectional AC Tracks (Bi) (i.e., signal bus lines)				Unidirectional AC Tracks (i.e., between PMOS and NMOS)			
P (µm)/N (µm)	(mA)	M1	M2	Contacts	Vias	M1	M2	Contacts	Vias
80/40	1.4	1.0 µm	1.2 µm	1	1	1.0 µm	1.2 µm	2	1
120/60	2.2	1.0 µm	1.2 µm	1	1	1.4 µm	1.2 µm	3	2
200/100	3.5	1.0 µm	1.2 µm	1	1	2.3 µm	1.8 µm	4	2
300/150	5.5	1.0 µm	1.2 µm	2	1	3.4 µm	2.7 µm	6	3
500/250	9	1.2 µm	1.2 µm	2	1	5.6 µm	4.5 µm	9	5
1,000/500	18.1	2.3 µm	1.8 µm	4	2	11.3 µm	9 µm	19	9

- If size is between the specified values, always approximate up

- 1.0 µm minimum Metal1 width

- 1.2 µm minimum Metal2 width

- Number of contacts and vias based on process current passage

I/O drivers use 3.3 V, but the internal core of the chip may run at 2.5 V to conserve power. In this case the chip may have two different voltages supplied from the PCB, or it may have an internal power supply circuitry that will regulate the external 3.3 V and provide 2.5 V to the chip core.

An extreme example of multiple power supplies is the design of DRAMs. A typical DRAM design requires an regulated internal VDD, two separate midpoint voltages, a super voltage known as VPP, I/O supplies VDDQ/VSSQ, and a negative substrate voltage known as VBB.

How do we effectively deal with such a large number of power supplies for issues such as routing and layout verification? Here are some suggestions and methodologies:

- Plan for each power supply in the floorplan and routing plan separately and in order of importance and complexity. For example, some supplies may be localized, such as the supply for an embedded memory, and therefore are lower in priority.

- Don't forget to plan for different substrate and well regions. For example, a negative VBB and VSS cannot connect directly to the same P+ substrate, so for a VBB substrate we have to plan for the added overhead of extra substrate connections.

- Power supply generators are tricky to floorplan, as they ideally should be close to the circuitry to which they supply power. This increases their effectiveness by reducing the required power supply grid to achieve reasonable parasitic resistances.

- Define all the powers appropriately in the command files for layout verification; otherwise, connectivity checks will not work. This includes definition of devices and substrate areas.
- Isolate large consumers of power with separate supply lines. I/O buffers are a good example, as most of the noise and surges will be isolated.
- Place reservoir capacitors for internal power supply generators close to the generators. These capacitors act essentially as batteries during high current consumption periods. Locating the reservoir capacitors away from the generator reduces the charging ability of the generator to the capacitors.
- Use clamping diodes to protect internal supplies from large power supply variations. Should the external power connection to the chip be oppositely applied, the VDD to VSS clamp diodes serve to limit the reverse voltage. As well, power supply diodes often form part of the overall chip ESD protection path.

An added feature of clamping diodes is the case where they are forward-biased when the difference between supplies exceeds one diode drop. In this way, current is "sourced" from both supplies when one becomes heavily loaded. Note, however, that this is not true for voltage dips on VDDQ, which should remain isolated from VDD during heavy loading.

Figure 7.28 illustrates a diode arrangement between separated core and pad supply pairs.

The layout of power supply diodes must take into consideration the large forward currents that exist during clamp conditions.

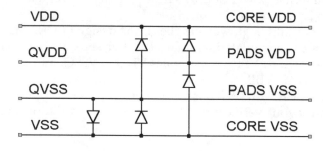

Figure 7.28 Power supply clamping diodes.

CHAPTER EIGHT

Layout Considerations due to Process Constraints

There are layout techniques that must be used to address limitations or effects to the circuitry as a result of the manufacturing process. Examples of these are described in this chapter.

8.1 WIDE METAL SLITS

Power supply lines in a chip are designed to be very wide so that electromigration and resistance effects are minimized. Are there maximum limits to the width of metal lines? In general the answer is no. However, there is one problem with very wide metal lines that occurs when the temperature of the chip rises high enough to cause the metal to expand significantly.

Figure 8.1 shows the effect of heat on the metal as it expands. As the metal heats up, the sideways inertia of a large piece of metal prevents sideways expansion. As a result the expansion of the metal is in the center. This causes the center areas of the metal to expand upwards. This effect is not as significant for smaller signals, as the upward expansion of the metal occurs with a smaller force because of the smaller size and lower sideways inertia.

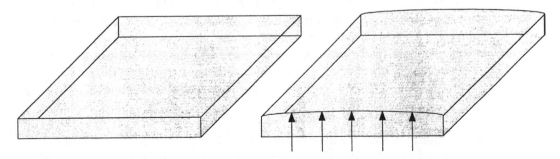

Figure 8.1 Metal expansion due to heat.

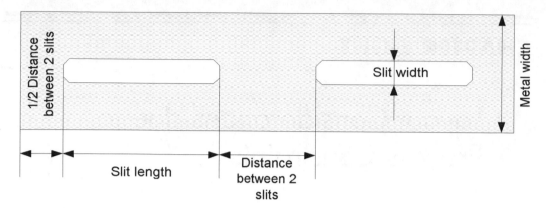

Figure 8.2 Metal slit example and design rules.

If the metal expands repeatedly with enough force, the metal will eventually crack the isolation and passivation layer that protect the wafer. Impurities and particles will work their way onto the chip, react with the different materials, and cause the chip to fail or work unreliably.

To address this problem, layout designers are required to put slits or holes in the metal at regular intervals. This technique has the effect of reducing a very wide metal to one that has many smaller areas that happen to be connected together.

Figure 8.2 shows an example of a metal line with metal slits cut out of it. The design rules for slits are very process dependent because they depend on the metal granularity, temperatures of expansion, type of material, etc.

Note the following:

- Slits have 45-degree corners to alleviate stress induced by high current densities within the metal.
- A general guideline for the maximum size of line that does not require a slit is 35 μm.
- Divide extra very wide metals into increments that are lower than the maximum allowed width. (For example, if the maximum metal width without slits is 50 μm and we are inserting slits into a 100-μm line, then it is advisable to use two slits in the line instead of one.)
- It is easiest to insert slits into a wide metal line by building a structurally correct unit cell and instantiating it as required.
- Slits should always be implemented in the direction of current flow. This is especially important for T junctions or other configurations.
- Slits generally can be implemented over a range of lengths. Try to use an average size so the slits can be easily adapted for special area such as corners and junctions.
- Discount the effective metal width of the line by the width of the slit or similarly add to the desired width of the line the width of a slit to account for

the lost area. For example, if a metal line is desired to be 100 μm wide, two slits are to be inserted, and the slit width is 5 μm, then the total width of the line will consume 110 μm of space.

- Analyze the current flow before you make slits in corners, T-junctions, and power pads.
- If the DRC cannot check all of the cases presented here, a visual inspection is necessary.

Design rules for the definition of slits are usually well described, but in many cases the conditions under which they are to be used are not. Specific areas where slits should be used are for power lines near the corners of the chip and the case of a pad connection in a T shape to a wide metal bus.

Figure 8.3 shows a corner power track and the way the design rules are to be respected in this case.

As we can see in Figure 8.3, it is important to have a small bridge of metal during the 45-degree turn to increase metal physical resistance against chip corner breakage during cut and package assembly. Figure 8.4 shows a proper connection from a power pad to an internal bus with correct metal slits. The current flow is shown as a guide for understanding the implementation of the slits.

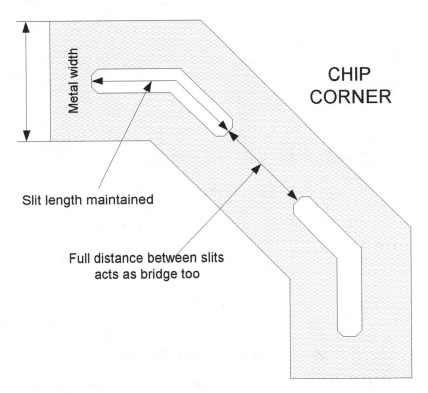

Figure 8.3 Corner routing power.

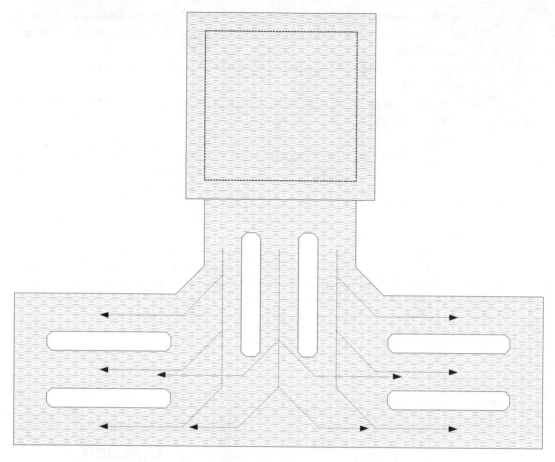

Figure 8.4 Pad connection with slits.

8.2 LARGE METAL VIA IMPLEMENTATIONS

As we have mentioned in Chapter 7, vias connecting layers together should be considered in the electromigration and resistance calculation. There are also process-related issues that should be considered.

The vias are structures that lie directly in the current path between the two layers. Thus, the layout design of interlayer connections using vias should be well understood. It is most important in large metal lines because in general it is these lines that carry large currents.

From a process point of view, vias are holes defined in the isolation layer between two layers: the top layer metal is required to fill the hole and connect to the lower layer. The manufacturing of this hole and the subsequent filling of this hole by the upper-layer metal does not result in connections that are the same size as drawn.

Figure 8.5 illustrates this effect. The design rules and electrical characteristics of the vias take these effects into account to ensure that a reliable via is formed. On the CD-ROM, there are more pictures of vias taken from a wafer.

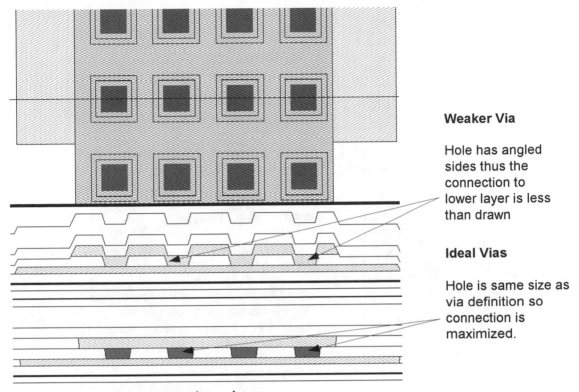

Weaker Via

Hole has angled
sides thus the
connection to
lower layer is less
than drawn

Ideal Vias

Hole is same size as
via definition so
connection is
maximized.

Figure 8.5 Cross-sectional view of via cuts.

There are techniques in layout design that may increase the reliability of via arrays. Common examples of these are shown in Figure 8.6, and their usefulness depends on the process.

The ability of a line to carry current is defined by the layers width and thickness. This is analogous to the size of a pipe carrying water. The number of vias connecting one layer to another must be determined by a variety of conditions, starting with electromigration, resistance of the via, current flows, process specifications, and planarization.

Via array configurations that are optimized for circuit performance are shown in Figure 8.7.

8.3   STEP COVERAGE RULES

For each type of design, ASIC or DRAM, for example, the processes are very different, as we have already explained. Based on the purpose of the chip, market prices, design requirements, etc., companies are developing special processes. The variety of design rules for each of these processes continues to evolve.

Layout designers are not involved in processing, but they have to take measures to prevent possible problems during the chemical and physical processing of the wafer. One problem that can be addressed with proper layout design techniques is the step coverage effect.

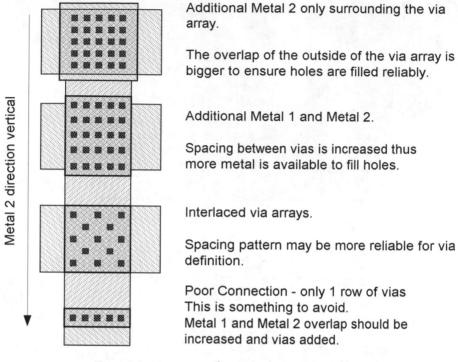

Additional Metal 2 only surrounding the via array.

The overlap of the outside of the via array is bigger to ensure holes are filled reliably.

Additional Metal 1 and Metal 2.

Spacing between vias is increased thus more metal is available to fill holes.

Interlaced via arrays.

Spacing pattern may be more reliable for via definition.

Poor Connection - only 1 row of vias
This is something to avoid.
Metal 1 and Metal 2 overlap should be increased and vias added.

Figure 8.6 Via array configurations for process reliability.

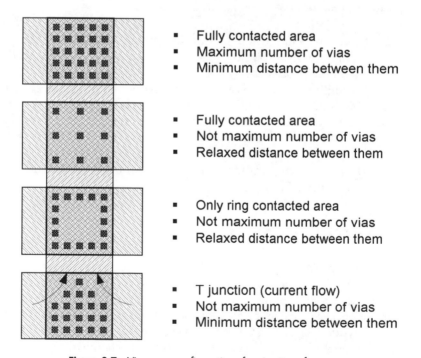

- Fully contacted area
- Maximum number of vias
- Minimum distance between them

- Fully contacted area
- Not maximum number of vias
- Relaxed distance between them

- Only ring contacted area
- Not maximum number of vias
- Relaxed distance between them

- T junction (current flow)
- Not maximum number of vias
- Minimum distance between them

Figure 8.7 Via array configurations for circuit performance.

A "step" in this context refers to the rising or falling slope of a layer as it passes between chip areas where a different number of layers exist underneath. For clarification, see a DRAM memory cell area versus the neighbors and the concept of friendly cells (Figure 5.23).

A process using a technique called planarization alleviates this problem, as the surface of the wafer is leveled with isolation material between layers. In this way, the steps are removed.

Why is this "step" a problem? If we analyze Figure 8.8, we can see that when metal1 is routed along a poly line, the angle defined by the height of the poly generates an irregularity in the metal cross-sectional shape. The layer is no longer the desired rectangle, but an odd shape that does not have the same characteristics as a straight line of metal running over a flat surface. In an extreme case, the metal line may physically break!

For example, if a via is placed on top of the irregular metal, the connection is likely to fail or be very unreliable. Figure 8.8 also shows the same situation for a planarized wafer, and the result is much better.

A design rule and layout requirement related to planarization is a rule defining specific density goals of a given layer over a specific area. For example, a rule might state that within areas of $100 \times 100\,\mu m^2$ regions, metal polygons must cover at least 75 percent of the area. The idea here is to implement enough polygons to ensure proper planarization for the layers above.

In order to meet this requirement, the layout may need to contain dummy or extra polygons to ensure that the process achieves layer consistency over the entire region. In some cases, designers can set up the layout verification tool to automatically find the density problems and fill the area with the dummy layer. These layers may electrically float, but in general they are connected to power supplies and add to their decoupling capacitance at the same time.

8.4   MULTIPLE RULE SETS

People working with ASIC processes generally have an easier time doing layout, as they have the benefits of having a relatively simple set of design rules. The total number of rules is less than the set of rules used by designers in DRAM or embedded memory processes.

Why do memory processes have significantly different rules from anything else? Because the memory cell fills 50 to 70 percent of the entire chip area, the memory cell itself is an extremely customized design with its own set of rules and specialized layers as well. As a reliability measure, memories have redundant circuits included on the die so failures in the array can be replaced.

Defining a final set of design rules is a very complex process in which circuit designers, layout experts, and process people have to trade off many factors: price, size, complexity, tolerances, design easiness, reliability of the process over the area of the chip and wafer, etc.

In terms of design rules, a DRAM memory has three different rule sets: one for the memory cells and their friendly cells, one for the pitch-related logic interfacing to the memory cells, and one for the periphery layout outside of the

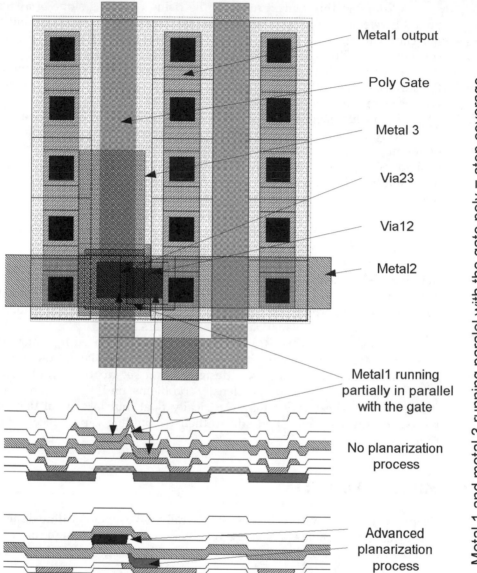

Figure 8.8 Step coverage problem.

memory array. The rules for the periphery layout may be similar to those for an ASIC process within the same process generation.

Table 8.1 shows an example of the definition of three rule sets for a DRAM process.

The layout of designs with multiple rule_sets must be done in such a way that the different set of rules can be verified using the DRC verification tool. The fundamental problem is identifying in the layout database areas that are to be checked with specific sets of design rules.

TABLE 8.1 Example of DRAM Design Rule Set

Design Rules for a DRAM Process	Cell	Pitch	Periphery
Active minimum width	0.5	0.5	0.6
Active to active distance	0.4	0.4	0.5
Minimum poly width	0.25	0.3	0.35
Minimum poly distance	0.3	0.35	0.4
Minimum metal width	0.4	0.4	0.5
Minimum metal distance	0.3	0.35	0.4

Three example methodologies are given below:

1. Each region defined by one particular set of rules is implemented using a separate set of layers. Masks are created from a database of merged layers once the layout verification is complete. This methodology requires the layout designers to switch layers in different regions, but once this is done, the regions are easily manipulated.
2. Each region is defined by a special layer called a blocking layer. General layout is done using the same layer in all regions, and once they are complete, the desired blocking layer is drawn. Within the layout verification tool, intermediate layers are generated to separate the polygons for each set of design rules as demonstrated in Table 8.2.
3. Regions are identified by a cell naming convention. The layout verification tools are set up to recognize and identify different regions by the name of a cell. In this case, the general layout is done with one set of layers, and once this is complete, the cell is named appropriately. This approach is simple; however, it relies on the fact that the layout can always be divided into discrete cells that do not overlap.

8.5 ANTENNA RULES

A side effect of the manufacturing process that leads to damaged parts is known as the antenna effect. Under certain conditions, plasma etchers or ion implanters induce charge onto various structures that connect to a gate of a transistor. The induced charge threatens to overstress and irreparably damage the thin gate oxides of the transistor, causing unreliable operation.

Charge is readily induced during the manufacturing process if a structure is built in such a way that it acts like an antenna. An example of an antenna structure is shown in Figure 8.9, where the ratio between poly over field (thick oxide)

TABLE 8.2 DRC Layer Generation Using Blocking Layer Strategy

Blocking Layers Merging Strategy

Active layer + MEM blocking layer =	Active memory
Active layer + PITCH blocking layer =	Active pitch
Active layer + PERY blocking layer =	Active pery
Poly layer + MEM blocking layer =	Poly memory
Poly layer + PITCH blocking layer =	Poly pitch
Poly layer + PERY blocking layer =	Poly pery

and poly over gate (thin oxide) is large enough. There are others specific to each process, and some include metal.

As the gate size gets smaller and more metals are added to a chip, and as process engineers reduce the thickness of the oxides, the antenna effect can have a greater impact on the yield of a wafer.

Approach 1 shows a technique to eliminate the antenna shown in the example by breaking up the long poly that acts like an antenna.

Approach 2 shows a diode placed near the transistor in danger to eliminate the effect of the metal antenna. As soon as enough charge is induced onto the metal by the antenna effect, the diode diverts the charge to the substrate.

In many advanced processes, the antenna rule checking is required for structures made of any number of routing layers, i.e., metal1, metal2, etc. Some of the automated routers know how to limit the routing in one metal for long distances in such a way that they can avoid the antenna problem.

Explicit and separate diodes attached to certain nodes may not be necessary if an inherent diode is attached to the line somewhere along the path. For example, a source or drain area of another transistor may be sufficient to act as a diode to divert unwanted charge.

8.6 SPECIAL DESIGN RULES

In previous chapters we learned the basics about process order flow and design rules related to process and circuit requirements. However, there are a few interesting exceptions to "general" design rules.

We cannot explain in detail all the "weird" rules related to specific types of processes. Our intent is to increase the reader's awareness of some process-related issues that are not always described from the beginning in the design rule sets. In general, it makes sense to talk to the process people to understand the technical reason behind any special design rules and try to work with them to determine practical solutions.

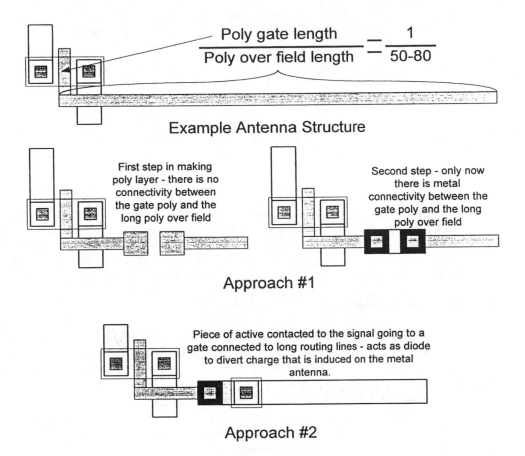

Figure 8.9   Antenna rules example and solution.

8.6.1 Minimum Area Rule

A rule that becomes more prevalent as geometry get smaller is the "minimum area" layer rule. Although design rules such as transistor gate length are shrinking, not all of the many layers in the manufacturing process shrink equally.

One example of this is the definition of active areas, especially for small polygons. This limitation is typically specified as a minimum width and area rule. Small active rectangles can occur quite frequently for substrate connections that are made with a single contact between metal1 and active. In this case, if only the minimum width rule is followed for the length and width of the polygon, an active polygon may result that is smaller than what can be produced.

Adding length to a polygon is an easy solution to conform to the "minimum area" rule. See Figure 8.10 for example layouts illustrating this rule.

8.6.2 End Overlap Rule

We learned about a generic contact overlap rule in Chapter 3, but in some processes this rule is enhanced by a rule known as the "end overlap" rule.

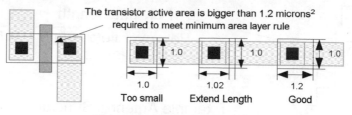

Minimum Area Rule = 1.2 microns²

Figure 8.10 Minimum area layer example.

This rule applies when a contact is located at the very end of a line. The problem that this rule addresses is shown in Figure 8.11 by the dashed lines denoted as "Shape on Silicon." Contacts that are placed at the very end of a line are in danger of not being filled as the metal is rounded in real silicon. This danger is increased if any amount of misalignment occurs during processing.

8.6.3 Double Contacts

Consider once again DRAM processes where the limits of the manufacturing process are consistently tested in order to achieve a small memory cell. In this case the overall reliability of many standard structures is in question for many design rules.

Outside of the memory array, where the topology of the layout is not regular, more stringent design rules are enforced. One of these is the requirement for double contact and vias for every connection. This rule applies to transistor layout and general routing and signal connections.

The double contact and/or double via not only improves the resistance of the connection; more importantly, it provides added reliability by having redundant contacts for every connection.

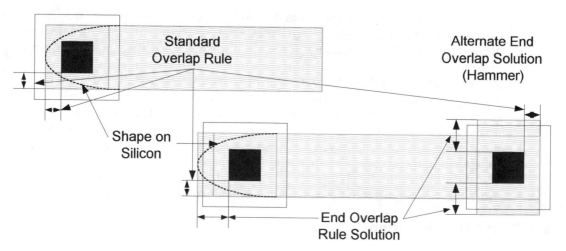

Figure 8.11 End overlap rule example.

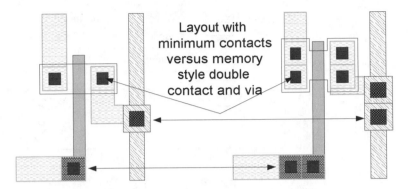

Figure 8.12 Single and double contact design styles.

A problem with this requirement is that placing double contacts using automatic tools is a challenge for place-and-route tools. These tools prefer square via/contact cuts (cells). Connections are easily made in both X and Y directions, and the tool does not have to orient the contact or via cell during placement.

Figure 8.12 shows two equivalent layout designs, one with single cut cells and the other with two. As we can see from the example, such layout styles increase chip size and the amount of work required to design a memory chip. For the present, DRAM layout designers can get very little help from the automation.

8.7   LATCH-UP

Conceptually, *latch-up* refers to the state of an IC when it is made inoperable by a parasitic shorting of VDD to VSS. Depending on the severity of the latch-up condition, the IC may be irreversibly damaged, or it may recover only after a complete power shutdown.

Let us try to understand, in a brief way, how latch-up comes about. Figure 8.13 shows the equivalent circuit model of the parasitic devices we refer to within a simple CMOS inverter. Transmission gates are a another risky source of latch-up, especially if the source or drain of the transmission gate is connected to VDD or VSS.

Redrawing the schematic of Figure 8.13 into a more readable format, we end up with the drawing shown in Figure 8.14. Very simply, latch-up occurs if either of the two bipolar transistors is turned on. If this happens, there is a positive feedback loop in that when one transistor turns on, the resulting current flow encourages the other transistor to turn on as well. Positive feedback occurs once again as the second transistor's current flow strengthens the first transistor's drive and the vicious cycle feeds upon itself. Under normal conditions, latch-up is not likely to occur, so how do these transistors become activated? First, let's very briefly discuss how bipolar transistors work.

Similar to MOS transistors, bipolar transistors are activated when there is a voltage difference between the base (labeled B) and the emitter (labeled E). NPN bipolar transistors require a positive VBE, whereas PNP transistors require a neg-

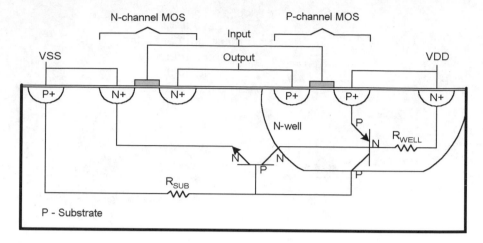

Figure 8.13 Inverter cross-section with latch-up circuit model.

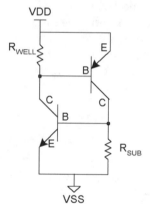

Figure 8.14 Redrawn latch-up circuit model.

ative VBE. We should recognize similar functionality of transistors to that of the voltage requirements in activating NMOS and PMOS transistors.

What triggers latch-up if under normal conditions these parasitic bipolar transistors are off? The most common trigger for latch-up is undesired or extreme currents injected into the Chip through the power supplies VDD or VSS. These nodes are connected directly to the outside world and thus are constantly exposed to uncertain voltage and current levels.

Let's go back to analyzing the schematic shown in Figure 8.14. As an example, let us imagine an abnormal current being injected into VDD from the outside world. This current results in a voltage drop across RWELL by Ohm's law—and voilè, we get our negative VBE for the PNP transistor, and it turns on. The current produced by the PNP transistor causes another voltage drop across RSUB, therefore turning on the NPN. The NPN transistor current sustains the voltage drop across RWELL, keeping the PNP transistor on even if the external VDD current has disappeared. The PNP current in turn sustains the operation of the NPN, and this vicious circle is latch-up.

In summary, here are the very simple requirements for latch-up to occur that we have discussed:

- A large enough V_{BE} generated (at least temporarily) to activate either bipolar transistor. This requires the combination of an abnormal current injected into the chip and parasitic resistance values large enough to generate this voltage difference by Ohm's law.
- Parasitic bipolar transistors of sufficient current drive strength to sustain the required V_{BE} to keep the bipolar transistors on.

Latch-up is a phenomenon that is well understood, as it has a long history in CMOS IC design. Many guidelines and design rules have been developed that inherently reduce the risk of latch-up and minimize the likelihood of meeting the requirements just listed.

A straightforward strategy to reduce the likelihood of latch-up would be to reduce the parasitic resistances across the base-emitter nodes and therefore increase the current requirement to trigger latch-up. If this is done properly, the trigger current for latch-up may be large enough that it is physically impossible.

In general, these techniques also weaken the strength of the bipolar transistors by breaking up the chip area into fragmented areas and therefore reduce the effective size of the various bipolar transistors.

Layout methodologies that reduce susceptibility to latch-up include the following:

- Avoid routing power supply lines (especially to substrate or tub contacts) in resistive materials such as diffusion or polysilicon. Keep the power nodes in metal!
- Place substrate and tub contacts between transistors of different types. In addition minimize the distance between substrate contacts and transistors within a well and vice versa. For example, if PMOS transistors are within an N-WELL then place the P-type substrate contacts as close as is allowed to the PMOS transistors. Apply the same logic to the N-type tub contact spacing to NMOS transistors.
- Maximize the number of substrate and tub contacts.
- Minimize the spacing between substrate and tub contacts.
- Ensure an even coverage of substrate and tub contacts over the entire area.
- Use continuous strips or bands of substrate and tub contacts. This technique is formally known as *guard banding*, especially when the bands completely surround transistor areas.
- Group transistors of the same type together to avoid the overhead of having to protect against latch-up in many different areas.
- Place internal circuitry away from external pad areas.
- Be extra careful in considering latch-up conditions in areas where the substrate or well is not the same potential as the source nodes of the transistors.

In all cases, it is always best to develop formal numerical design rules that can be checked by the layout verification tools. Many of the methodologies just

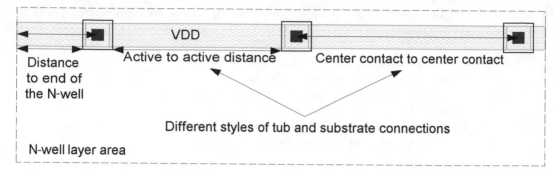

Figure 8.15 Tub connecting the N-well to VDD voltage.

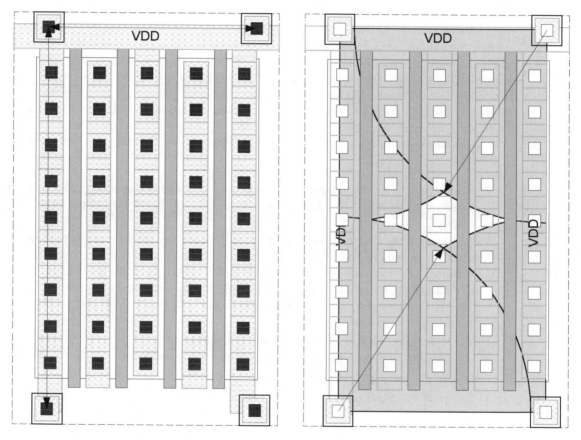

Figure 8.16 Example of tub area coverage inside N-well.

listed are not easily converted to numerical checks, but we recommend at least the following: A design rule specifying a maximum distance between substrate or tub contacts.

In general this distance is about 40 to 50 μm, but as the gate length of transistors shrinks, this guideline should be reduced. This rule ensures an even coverage of a given area with bulk connections, and there will be consistent voltage all over the area.

There are two basic ways of checking this distance: from active to active or from contact to contact. Figure 8.15 shows both ways of measuring these distances. Both styles are correct depending on the number used to check the distance between them.

The second rule is area coverage of each contact based on substrate or N-well resistance. Each of these actives has substrate coverage to prevent latch-up of about half the minimum distance between them. So if the minimum distance is $50\mu m$, then the coverage is $25\mu m$.

Note from Figure 8.16 that while the layout conforms to the rule just given, the distance between the four contacts leaves a small island of transistor area that is not protected. The shaded area from each contact (gray) with the dark lines is the coverage radius of the four tub contacts. The noncovered area is the middle white region.

For a layout to be completely verified, both of these rules need to be satisfied.

Big transistors or very high-speed circuits that are switching fast and continuously can inject a lot of noise into the substrate by their size. Examples are clock generators or output buffers. It is advisable to use full guard rings for all tubs and substrate connection around the transistors.

In layout, these guard rings are divided in two basic kinds, hard and soft ties. The hard tie is a fully contacted ring of active that has continuous metal1 over it. Soft ties have a continuous ring of active, but the metal layer may be broken to accommodate signals passing into the guard ring.

Figure 8.17 illustrates the differences between hard and soft ties.

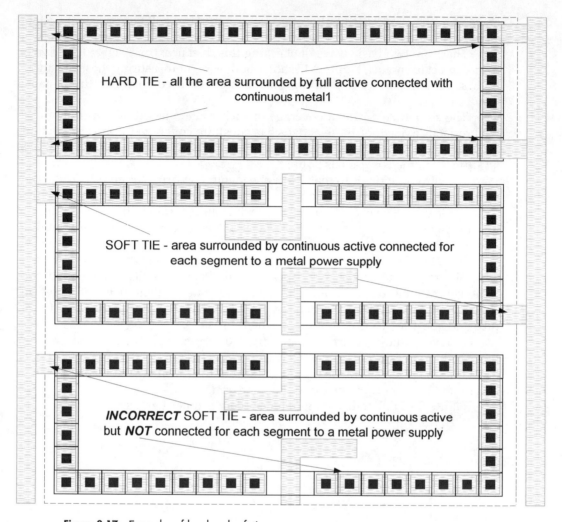

Figure 8.17 Examples of hard and soft ties.

CHAPTER NINE

Layout Design Techniques in an Uncertain Environment

There are many layout techniques that can be used to anticipate change to the circuit design. Last-minute changes are common to all projects, as there is never enough time to fully verify a design. With this mentality of anticipating change, techniques that minimize the impact on the overall chip layout are described in this chapter.

9.1 LAYOUT OF CIRCUITS DESIGNED FOR CHANGE

Although changes to circuit designs are impossible to forecast, there are many cases where we can predict that some circuits will require tweaking. In this case structures to adjust the functionality or timing of a circuit are included in the schematic from the start. There are preferred layout techniques that are used to implement these circuit designs.

9.1.1 Metal Option Programmability

Metal option programmability is a methodology that is similar to the idea behind gate array design, but applied in very specific applications. The idea is to have a design that has a common set of base circuitry and layout and is "programmed" by different configurations of metal masks.

In the context of the layout of circuits designed for change, metal option programmability is the application of this concept to individual circuits. Gate arrays have the concept of a master slice where the entire die is programmable. Metal options refer to a specific design such as a delay circuit, control logic, or memory configuration that can be modified or customized in a much smaller way.

The applications of metal options are virtually endless and perhaps limited only by the designer's imagination. However, examples where metal options are used and are very valuable include the following:

- Reconfiguration of the functionality of a design. In this case, different metal options would produce different end products from a design with a common set of base layers. The following are examples of features that can be reconfigured:

  Operational features such as low power modes and test modes

  I/O interface standard: TTL, LVTTL, SSTL

  I/O data width: ×1, ×4, ×8, ×16, ×9, ×18

  Power supply voltage: 1.8 V, 2.5 V, 3.3 V, 5 V

- Analog circuit fine-tuning such as resistor values or devices sizes. The example of an ESD circuit adjustment was discussed in Chapter 6.

- Evaluating new or unproven circuitry against an established design. For example, two circuit designs may be implemented on a chip and a metal option might be used to switch between them to compare one with the other. This is most useful for tricky analog circuits such as PLLs, input buffers, and oscillators.

- Circuit adjustment by switching in or out devices, cells, or logic gates.

Let's examine in detail an example that is very common. Delay chains are one of the cases where metal options are beneficial, since they are used to solve many different and sensitive circuit design issues.

Figure 9.1 shows different versions of one schematic delay chain that is configurable between 0 and 3 delay stages.

The schematics show CLOSED or OPEN switches that correspond to short and open circuits. The dashed line denotes the delay path in each case.

How are these options implemented and used in the design flow?

1. A circuit designer provides a drawn schematic with the options correctly drawn.

2. Two additional layers are defined, corresponding to the OPEN and CLOSED switches. These layers must be associated with a mask layer in which the options will be implemented.

For example, it is typical to implement options in top-layer metal for a two-layer metal process; thus, the options would be associated with metal2.

Using separate layers and devices allows the layout verification tools to verify connectivity and ensure that the options follow established guidelines. These guidelines should reflect requirements for DRC and product test where these option areas can be used once again.

The drawing and DRC checking of the options must be implemented carefully. CLOSED layers can at any time change to OPEN, so the layout design and checking should accommodate this flexibility.

Polygons drawn in these layers are considered devices similar to resistors and must be connected as shown in the schematic. An example layout is shown in Figure 9.2.

A design with metal options is a case where the layout can begin before the final configuration of options has been finalized. Once the layout has been completed, it is an extremely simple matter to change options from one to another, since the area and topology of the cell is not affected by a change in option layer.

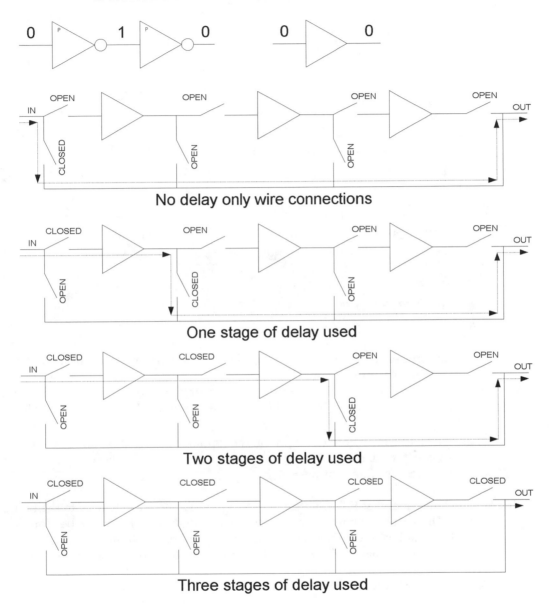

Figure 9.1 Example of a configurable delay chain.

1. When generating the final GDSII for tape-out or mask-making, the CLOSED layer is merged with the associated mask layer (in this case metal2) and the OPEN layer is discarded.
2. During prototype testing, the product test engineer can revise a silicon version of the circuit. One option is to use a focused ion beam (FIB) machine to alter the circuit once more. CLOSED option points can be opened or OPEN

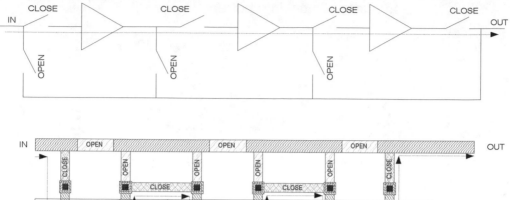

Figure 9.2 Delay chain layout.

option points can be closed using this machine. Another option is to use a laser to cut CLOSED options.

The guidelines for option layers mentioned earlier need to include limitations imposed by the method that is intended to be used on silicon prototypes.

3. Once the final configuration of the options has been established using silicon-proven results, the schematic and layout database need to be updated to reflect the final settings.

Remember that FIB modifications to a particular die apply only to the FIBed chips. Therefore, if changes to the masks are required, this must be done by updating the layout database and regenerating the required masks.

Production masks are produced from this updated database.

9.1.2 Via Programmability

In the context of the layout of circuits designed for change, via programmability applies to circuitry that is truly designed to be programmed, such as read-only memories (ROMs) or specialized decoders that are configured independently from the base design.

In this case, the configuration of a circuit is driven by a "coding" scheme that is produced in general from a software development group. Implementing the coding scheme is done by the placement of vias on top of a programmable circuit.

There are two aspects to implementing a layout design for via programmability:

1. *An unconfigured layout design that can accommodate all possible placements of vias.* "Legal" sites should be designed into the leaf cells and assembled design. Legal sites for via placement are defined as locations where the resulting design would be design-rule correct under all conditions.

2. *A methodology for placing or programming the vias onto the base layout design.* This is usually a macro that uses a polygon editor as the layout engine, or a customized CAD package that can read and produce a layout database in a standard format.

Via programmability using software is a common technique used in ROM design today, and it has the following advantages:

- Software designers for "on the chip" can optimize their code until late in the project schedule.
- If planned and implemented properly, the design is correct by construction and should always be DRC and LVS error free.
- Once the via programming technology is developed, it can be reused for subsequent evolutions in the underlying base layout. Only the "legal sites" of a new architecture need to be specified.
- An effective base layout design and programming scheme can include techniques to minimize the connectivity of devices by excluding vias in the programmed design. Reductions in power consumption and increase in speed result.

Figure 9.3 illustrates an example of via programmability for a small decoder used within MOSAID.

9.1.3 Test and Probe Pads

Test and probe pads are aids that are implemented into a design to enable product engineers to check the internal operation of a physical chip.

Test pads refer to pads of metal that are compatible for bonding; *probe pads* are much smaller and are intended for very specialized measurement devices.

Test pads are also defined to be only available on a wafer and are not intended to be bonded for the final product. These pads are uncovered by the passivation for the prototyping or evaluation phase and are covered by a glass isolation layer before it is packaged. In terms of ESD, they are less protected than regular pads because after testing they will not be connected to the pins of the package.

An example where test pads are used extensively is within DRAM designs that have internally generated power supplies for internal circuitry. The test pads are normal-size pads connected to internal supply powers such as VPP, VBB, VCP, and VBLP. During the prototyping phase, the test pads are used to determine if the internal power supply circuitry is functional. They are also used in production test to evaluate reliability and process characteristics of the memory over the life of the product.

Probe pads, on the other hand, are used exclusively for the preproduction debugging and evaluation stage. Product engineers use probe pads to check the

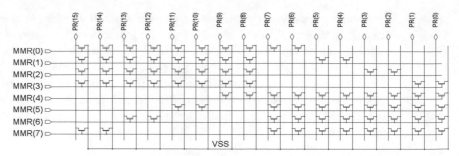

Fixed symmetrical structure having on each PR or MMR line
the same total number of transistors

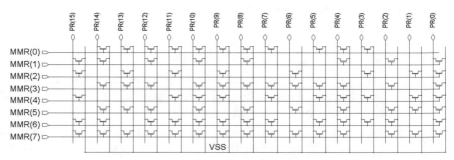

Total random transistors connections having on PR or MMR
lines a different number of transistors

Figure 9.3   Decoder architectures.

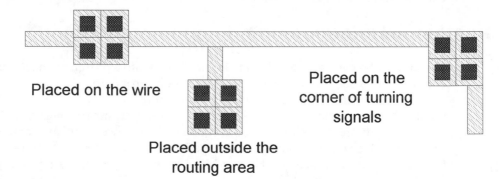

Placed on the wire

Placed on the
corner of turning
signals

Placed outside the
routing area

Figure 9.4   Probe pad examples.

functionality and timing of internal nodes of chips that have not been qualified
for production.

Probe pads can be placed anywhere on the die, since they are used when
the die is completely unprotected (Figure 9.4). The size of the probe pads is
dependent on the equipment that is used.

Historically, probe pads were not explicitly implemented, because a single
via was big enough for the probes that were available. As layer geometries have
shrunk, a single via has become too small for physical probing. Generally, a probe
pad that is in the range of $1\,\mu m^2$ is sufficient.

Probe pads should be implemented in layout as directed by the circuit designer. However, it is likely that almost all routed nodes should be probeable. Certainly, critical signals such as clock signals and datapath signals are prime candidates for probe pads.

It is important to document the probe pads coordinates carefully so that they can be easily found!

9.2 PLANNING FOR UNKNOWN CHANGES

During various phases of design, layout designers are challenged with design rule changes, circuit changes, bugs in the software tools, etc. The best medication is prevention, but we don't always know what to expect. In a DRAM chip, if any of the major memory layers gets a minor design rule change, it may not be efficient to use the planned architecture, so the entire core layout and design should change. Such process changes have much less impact on the layout of ASIC designs, because their processes are much simpler and because most of the layout is done using place-and-route tools.

Planning for unknown changes might seem strange at first, but we can always estimate the amount of change of a particular design based on the novelty of the design and the experience of the circuit designers. In general, when planning at the chip level, most numbers are "wet finger" values. Even using the most advanced tools for floorplanning, there is no way to envision the final architecture, the final number of gates, or the total number of signals, for that matter. The best solution is to plan the chip with change in mind, and to forecast the type and amount of change based on personal experience.

As an example, if a design is estimated to have 50 percent of the circuitry that is completely new, we can start to plan on 50 to 60 percent of contingency in our work. With this in mind, spare logic, area, signals, etc., can be reserved to cope with design development and changes. Using this approach will give the final design a much better chance to end up a success.

Conversely, if contingency planning is not done, the size and schedule of the initial design will grow and grow to the point where the project will be at risk. We can tell you a secret: "everything is possible in layout" where only the imagination is the limit, but more and more, "time to market" is the real limitation. Preparing a plan that can accommodate changes at any stage with minimum schedule changes and almost no change in chip size is the dream of all project leaders today. Plan for change when starting any part of the design, because people make mistakes, and the dream can become reality.

Last-minute changes are handled using a formal procedure called an Engineering Change Order (ECO), described in the next section. Before we receive and have to act on one of these, there are many methodologies we can implement that will help us handle these emergencies. This section outlines a few of them.

9.2.1 Contact and Via Instances

The concept of a hierarchical design was described in detail in Chapter 3. There are many benefits of using hierarchy, and in the context of planning for change, the use of hierarchy is key to minimizing database management issues when reacting to large-scale changes.

This section is devoted to discussing the merits of contact and via cells specifically because there are many benefits to them. Place-and-route tools use them for connections, but the concept was developed before the advent of these tools.

The idea was to develop a library of cells for various types of contacts. This practice is effective when there are three to five types of contacts, but in the case of DRAM memories in particular, this concept really improves layout efficiency. DRAM memories have many poly layers, but also have multiple design rule sets, so there is a great range of contact and via cells. It is not uncommon to have a library of two to three dozen cells.

Historically, contact cells contained a single polygon: the contact layer alone. The idea that we are proposing (within MOSAID it is standard procedure) is to create the contact cells including all three related layers. For example, a contact between active and metal1 will be a cell that contains contact layer, active, and metal1 overlapping the contact. The layout conforms to the design rules completely so that the contact "cell" is DRC clean.

Some advantages of this approach include the following:

- Advanced rules such as the "metal end overlap" rule can be implemented globally without too much effort.
- The cells are DRC clean, so other designers will create layout that is correct by construction. The idea is that the contact cells are by definition correct, as with standard cell libraries for an ASIC designer.
- Defining naming conventions for each chip with multiple design rule sets will help layout designers to use the appropriate cell. Also, they do not have to memorize the different contact design rules—only those for the region they are working in.
- Cell names are portable from project to project. Memorizing numerical design rules is not.
- Contact cells are easy to connect to, since touching or overlapping the boundary of the cell is sufficient.
- The biggest advantage is when the time to tape-out is critical. In "bleeding edge technologies" such as memory designs, when people start the layout of a chip, the process is not yet fully defined. Many times in my experience, this solution saved the tape-out. In many cases we had to change the contact size, the overlap layers rules, or just the contact layer. Under normal conditions, all these changes will require some CAD software "processing," but such actions have to be checked and debugged before they are applied to a full chip. Using contact cells, it takes about 5 minutes to complete a global process change, with guaranteed success in our experience.

Because we found this feature very efficient, we decided to use it everywhere. We are trying to convince all the vendors offering transistor-level automation to

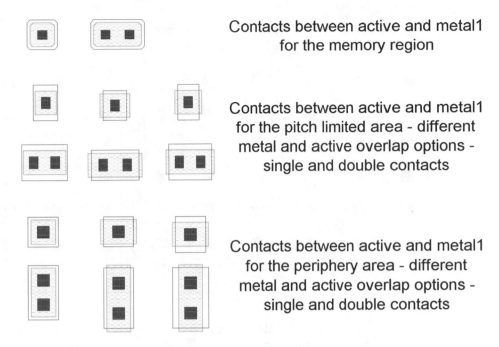

Contacts between active and metal1
for the memory region

Contacts between active and metal1
for the pitch limited area - different
metal and active overlap options -
single and double contacts

Contacts between active and metal1
for the periphery area - different
metal and active overlap options -
single and double contacts

Figure 9.5 Contact examples.

embrace the contact cell as a feature in their tools. In Figure 9.5 we can observe some examples of contact cells, including all their overlapping layers.

9.2.2 Minimum Design Rules?

Another approach to accommodating change and preventing major impacts on the layout design is to avoid using minimum design rules or minimum area design methods all of the time. This is especially valuable if it is known that the design under development is new or has a high probability of change.

Consider the scenario of an analog design, where the designer may not know from the start the internal routing scheme of the block. In this case the final sizes within his or her design will vary based on the final results obtained from layout extraction.

How can a layout designer prevent moving contacts, devices, and cells to accommodate anticipated changes? In this section, we outline a solution.

We learned in previous chapters that one of the important steps in layout design is to try to use source/drain sharing actives as much as possible to reduce the size of the area. In case of "expected" changes, the layout should be built in such a way that it could accommodate limited size changes. Figure 9.6 illustrates a technique that accomplishes this.

One side of the empty channel is power connected, while the other is an output. This is the output of a transistor that is likely to change.

The technique that is used here is to space the active regions by the amount that accommodates a finger of gate poly. The spacing may not be minimum, but allows significant flexibility in changing the transistor size.

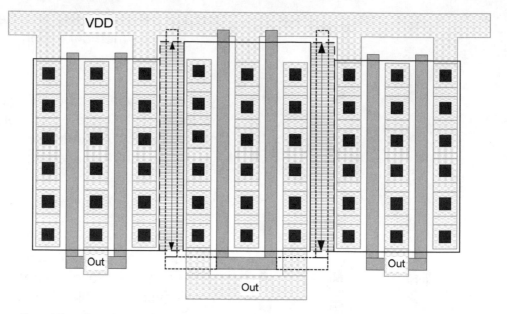

By adding 2 active and 2 gate polygons the center transistor can increase in
size by 2 fingers without any implications on the cell or chip size

Figure 9.6 Drivers size adjustments.

9.2.3 Spare Logic and Spare Lines

Even if the most advanced tools and experts are used, a chip will always have
some dead space central enough to make the idea of spare logic gates and
spare lines worthwhile. Unless the chip has a flat netlist and the design team
is using an automated place-and-route tool, any design will end up with holes of
unused area.

After expending the effort needed to implement last-minute changes and
after feeling the pain of bug fixes and mask revisions to revise only a small portion
of a design, many designers have accepted the idea of placing unused or spare
gates and signals on the chip before tape-out.

A simple technique is to devise a block of logic that comprises a common
group of logical functions and place the cell in any free area in the chip design.
The quantity and types of gates that are used may vary from chip to chip based
on a forecast of what may fail. The design of the block is heavily metal oriented
so that connections and reconfiguration of the transistors are easily done using a
minimum number of layers. The gates are disabled initially in such a way that
they do not affect the normal operation of the chip.

Combined with spare lines that connect various regions of the chip and
passing by the groups of spare logic, fixes to minor bugs become much easier.
Additional lines are not a great impact on chip size when compared to missing a
market window.

Figure 9.7 illustrates this concept of inserting spare logic into empty areas
that have good access to spare lines. Spare lines are routed in every channel, and

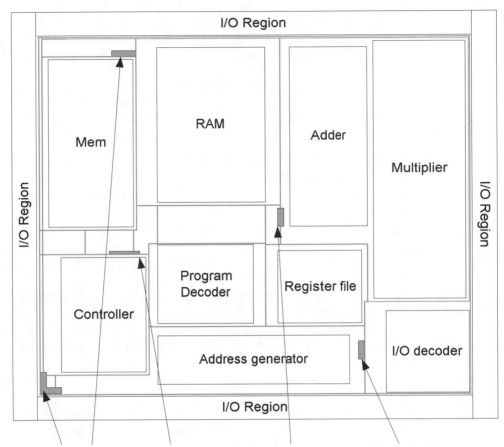

Small blocks of spare logic with access to a lot of spare lines

Figure 9.7 Spare logic and spare lines.

in general up to 5 percent of the number of signals in the channel should be spare not including routing requirements for the chip finishing stage. During the prototyping stage, it is a good practice to have at least two spare lines per channel. The spare lines respect channel metal directions and are fully connected from one end of the chip to another. Using numerous local spare structures, global changes can be implemented without any chip size or schedule impact.

9.3 ENGINEERING CHANGE ORDERS

Engineering Change Orders or ECOs are part of being a layout designer. As a design nears completion, verifying and integrating changes becomes a significant problem. Using a controlled methodology, such as an ECO flow, to implement change ensures that the bug is reviewed from a broad level of architecture and all the possible implications of the change in one block are reviewed against all the other blocks' interface points of view.

Engineering Change Orders are a formal methodology used toward the end of design projects so that changes are carefully implemented without delaying the project further. Using all of the methodologies that were presented, most changes would have minimal impact on layout design.

In terms of a procedure for implementing an ECO, there are as many ways of defining it as there are design flows and project management styles.

Conceptually, or from a management standpoint, an ECO procedure might look like the following:

1. The design is frozen on a given date with the understanding that full chip simulation and verification is ongoing.
2. At this point the layout may be in the last stages of finishing major blocks and the routing of the top level is 90 percent complete.
3. Any change from this point on is considered an ECO. The ECO is reviewed from a technical point of view to ensure that it is valid, and a ranking of its importance is determined. The impact on the project schedule is also estimated.
4. The project leader or manager reviews the ECO and decides whether it is to be implemented. ECOs are often rejected if the error is minor or can be corrected in a future version of the chip. Another criterion for evaluating an ECO is its relevance to the market or customer or the impact of delaying the chip's entry into the marketplace.
5. If the ECO is to be implemented and shortcuts are necessary for it to be done on time, then special precautions may be taken to ensure that the ECO is done correctly. Examples might be having a second person inspect the implementation, or a staged release of the design.

In the case of an ASIC flow where place-and-route tools are used to generate the layout of the chip, the layout tools have an ECO flow built into the tools.

The ECO procedure outlined below is conceptually straightforward, but not always so in reality:

1. The new netlist is reviewed and an approach to address the change is generated.
2. The place-and-route tools are deleting cells that no longer exist in the netlist and try to place the new ones. In some cases, if the design was very crowded and there is not too much free space, the new cells may be placed far away from the previously deleted ones, redoing the cell placement. The problem is that in many cases the initial placement was optimized for power, electromigration, RC, etc., and the ECO may change the picture.
3. The router is ripping up the local routing and is trying to reroute the new connectivity. Most of the popular routers today can do it in 99 percent of the cases. If the router gives up, human eye and manual edits can finish the last 1 percent of the job.
4. After the place-and-route is done, an extraction of the layout is required to check if the changes achieved the timing goals and there is not too much change to the related circuitry. If not, additional iterations of the ECO flow are needed.

9.4   GUIDELINES FOR PROPER LAYOUT

We described in previous chapters most of the basic layout concepts and methodologies, and now it is the time to summarize them in a consolidated list. The list is built based on the basic flow of top-down planning and bottom-up execution in design.

9.4.1 Chip Floorplan

- Learn the architecture of the chip from the designer, who has a vision of the possible blocks, power requirements, groups of signals, and new areas that were never designed in the company before.
- It may be possible to extrapolate from previous projects the size of some blocks and the number of signals related to them. We need this information to assess contingency for the chip and for each individual block.

If new types of blocks are to be developed, new flow and new tools may have to be brought in. If so, the project may need additional CAD support to introduce the new technologies. Place-and-route, compactors, new verification tools, libraries from another vendor, and IP blocks are only a few example of new technologies that may have to be included in the plan.

- Assess and compare process rules related to other known processes. The list includes the following:
  Vertical connectivity diagram—especially in case of multiple poly types
  Resistance and capacitance of gate, metal conductors, vias, and contacts
  Electromigration values for each conductor layer
- Evaluate the routing layers and determine the routing grid for each layer. Determine whether it makes sense to unify the routing grids to a single value for each direction for simplicity.

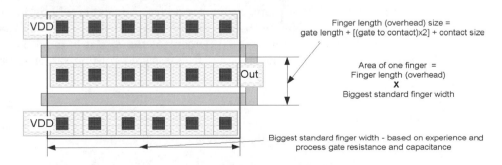

Block size = Number of transistors x Area of one finger

Figure 9.8 Basic calculation scheme for floorplanning.

- Calculate block size based on finger size (Figure 9.8) and add contingency based on the novelty of the design and the experience of the designer.
- Check package and define pad locations, especially for power and busses of addresses and/or data.
- Define power style and grid—power line widths based on electromigration or RC, and where the chip needs special power lines connections in order to achieve the allowed IR drop and resistance. In some advanced processes, power lines have their own routing layers.
- Evaluate the total number of signals; build channels including contingency and plan for spare lines and spare logic.
- Find repeated structures and try to reuse any possible piece to increase efficiency and reduce change impacts.
- Plan layout hierarchy and design to match each other as much as possible (same level and names). Even though the verification tools are fast today, debugging time is based on the user's ability to understand the errors and fix them in a timely manner.
- Update the floorplan regularly by replacing the empty boxes with finished blocks and rerouting the interblock connectivity. This way, when all the blocks are done, the chip is complete.

9.4.2 Blocks

Block-level layout guidelines are as follows:

- Import from the chip level, the block size and port positions and try to stick to them.
- Define feed-through signals so the overall chip congestion inside interblock channels will be reduced.
- Define power needs and grid inside each block.
- Define critical path design and group the related cells to minimize routing.
- Plan for changes by having spare logic and lines.
- Try to improve efficiency by using automated tools. When the number of components is 100 or more, even a good layout designer will find it hard to respect all the design constraints for this many instances and still be fast. Even if the tool is providing 80 percent of the job, it is much faster and less prone to error if the designer uses assisted routing rather than hand connectivity.

9.4.3 Cells

For cell-level design, we offer the following overall guidelines:

- Define cell architecture—i.e., standard, full custom, datapath, etc.—and boundary rules based on the block plan.
- Define power style and widths.

- Define special requirements—i.e., symmetry, neighbors, critical path.
- Define transistor style based on speed, power consumption, and routing requirements (porosity).
- Verify the cell DRC together with neighbors. This eliminates all possibility of finding errors at a higher level of hierarchy.

CHAPTER TEN

Computer-Aided Design (CAD) Tools for Layout

10.1 INTRODUCTION

Every year at the biggest conference for IC CAD known as the Design Automation Conference, VLSI designers from around the world are bombarded with names of new products and methodologies that can improve the efficiency of design.

Before the conference, magazine and e-mail advertising is frenzied, and during the conference, presentations and demonstrations are given all day long. In all of them the marketing and salespeople presenting the tools always feature new options and benefits such as extra capacity and quality, a variety of menu options, and the very "important" price/performance ratio in comparison to manual methods or tools from the competition. None of these presentations discuss the philosophy of the tool, what the designer of the tool had in mind, what new concepts the tool is addressing, or what kind of flow the tool is supporting.

We learn about buttons, ease of use features, values to enter into forms, but not the real reason why we should use the tool. In many cases, new tools that have good potential are not successful because they were not introduced with the right approach.

The vendors who market their tools to the right users are always more successful than others who don't, even if those others have a better tool. During the past 15 years many companies tried to build tools to totally eliminate the work of a layout designer by offering a "push-button" solution. However, most of them have disappeared.

As an example, while working at Motorola we received a demo of a new tool for layout design. The application engineer who came to present the tool started the demo with the following statement: "This is a tool that will provide an engineer with everything he needs to avoid using the services of a layout designer!" I don't think that I have to explain how successful the demo was and how a wrong marketing motto helped a very capable tool to fail.

Their timing was good, the tool was very capable at the time, and with some good customers and an installed base, it could have become the market leader. They were the first to see the advantage of integration between layout and circuit simulation databases. In this case the developers addressed the needs of one group of users by creating a disservice to the other. For obvious reasons, we will not give you the name of that tool in this book.

In this chapter we try to cover for each family of tools the concepts that we understand they are addressing and the methodologies required to use the tools. For a comprehensive list of vendors and tools, check out the Integrated System Design Magazine Web site, where there is always an updated vendor list.

A description of the basic types of tools used for layout today in CMOS VLSI designs is outlined next. Subsequent sections will expand on features and ideal usage of specific tools based on the flow and the context where they are or can be used.

The various classes of tools have been grouped into three areas, as shown in Figure 10.1.

Planning tools include the following:

- *Floorplanners:* A floorplanner is used to coordinate placement and routing engines to create a layout floorplan. Floorplanners are discussed extensively because they pose new concepts and challenges for layout design. When used properly, floorplanners can reduce time to market by provid-

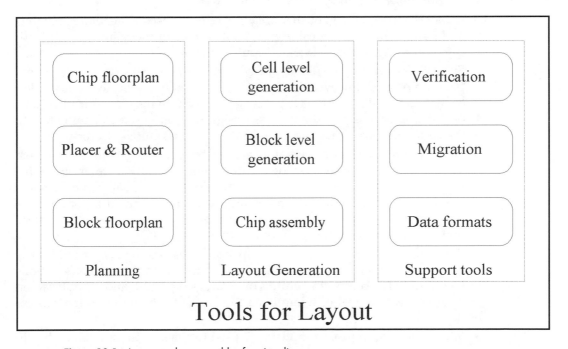

Figure 10.1 Layout tools grouped by functionality.

ing a methodology for top-down design and layout that is correct by construction.

- *Placers:* A placer optimizes the placement of cells or devices using physical and logical constraints. Placers are generally designed to work with specific routers; therefore, it is very important to use a placer and router from the same vendor. This is because the two tools work together to meet the constraints and take advantage of features and information that are known to both tools.

  To understand more about this topic, we will expand on the features and constraints of placers in the sections related to specific levels of design. Once understood, it should be clearer why it is better for a placer and router to work together. As a simple example, a channel router needs channels to route with, so the placer must provide these channels. If not, the world's best placer may be useless.

  In terms of types, there are three basic levels of placers, each having different features and requirements:

  Inside cells—transistor and cell placement

  Inside blocks—cells, blocks, or mixed cell and block placement

  Chip level—block-level placement within a floorplanning tool

- *Routers:* Routers were the first automation tools that were widely used. A router enhances the speed of layout interconnect. At first routers were capable of chip-level routing; they have evolved to handle cell-level routing today. A router is a must when the complexity of connections is beyond the capabilities and efficiency of a manual approach.

  Layout generation tools include the following:

- *Layout editor:* A polygon pusher or layout editor is used to generate polygons and paths using a graphical user interface. Some of them are very sophisticated and may include place-and-route functionality.

- *Symbolic editor:* A symbolic layout editor has the same user interface as a polygon pusher. However, the layout is generated symbolically from a coded or mathematical algorithm that is programmed into the tool. The advantage of this approach is that the process design rules are used as parameters to the code; therefore, it is easy to generate layout for different processes.

- *Device generators:* Device generators are used to generate layout devices such as transistors, via arrays, or logic gates. They typically have an extensive graphical user interface and a highly developed macro language. In some cases the device generator is an enhancement to a layout editor or an independent tool. Without a placer or a router, device generators have very limited value for enhancing productivity.

- *Compactors:* A compactor automatically optimizes existing layout and is generally used as an enhancement to an advanced layout editor or symbolic layout tool. The compactor shrinks or enlarges the width and space between polygons with a goal of minimizing the layout to the limits of the process design rules.

- *Silicon compilers:* Silicon compilers are used to generate layout automatically by generating transistors, leaf cells and structures using the leaf cells based on a standard architecture. In general, silicon compilers do not have a graphical user interface, as they are used to process a large number of structures. They are developed mostly by and for people with a lot of software experience.

Finally, the layout support tools include the following:

- *Layout verification tools:* Layout verification tools perform a suite of tests on completed layout. Design Rules Checkers (DRC), Layout versus Schematics (LVS), Electrical Rules Checkers (ERC), Layout Parasitic Extraction (LPE), and optical proximity tools are discussed.
- *Plotters:* Plotters are generally not a software tool, but in order to produce meaningful and readable hard copy of layout, software specifically designed to process layout is normally used.
- *Converters:* Layout converters are often called migration tools. Similar in concept to compactors, a converter is used to retarget or compact a previous design into a process with different design rules.

Finally, a discussion of data formats for layout databases will be presented.

10.2 PLANNING TOOLS

There needs to be a planning phase for all layout design tasks, but in general a floorplan starts from the chip level down to the block level. The idea behind this methodology is to build everything bottom-up while using the top-level floorplan to define the block interfaces and to coordinate updates to these interfaces as portions of the design are completed or verified.

10.2.1 Chip Floorplanning Tools

Floorplanning is the process of identifying structures that should be placed together and allocating space for them so as to meet the conflicting goals of available space (cost of the chip), required performance, and the desire to have every block connect seamlessly to everything else.

In most chips, the smallest design is also the highest performance design. Therefore, area and speed are characteristics that go hand-in-hand. A block or chip that is small in area has shorter interconnect lines, less routing, faster end-to-end signal paths, and even faster and more consistent place-and-route times.

Floorplanning is methodology that should result in a smaller design because the design is planned efficiently by combining the expertise of the layout designer in partitioning the circuitry and the optimization algorithms in the floorplanning tool.

Figure 10.2 shows the different components of a floorplan at the start of the design process.

To understand the role of a floorplan and how a tool can help the designer to obtain fast and correct results, it is important to understand the concepts and issues that a floorplanning tool is trying to address. A floorplanning tool does the following:

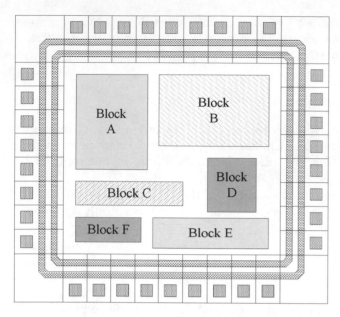

- Boundary and ports of the chip with or without pads

- Blocks and their internal pins

- Channels for (maybe) required routing

Figure 10.2 Components of a simple chip floorplan.

- Understands the different partitions or blocks of a design
- Understands the critical characteristics of each of the blocks: size, aspect ratio, and pins
- Dynamically displays the connectivity between blocks and connections to the pads
- Allocates space for routing based on number of routing layers
- Places top-level ports based on constraints
- Places each block and optimizes the pin locations for each block based on the overall connectivity requirements and the feasibility of routing the signals between blocks
- Allows the user to make incremental modifications to the plan or to replan altogether

Figure 10.3 highlights the important information on a block that is required before it can be instantiated in the chip-level floorplan.

Figure 10.4 outlines a procedure for using a floorplanner effectively. Here are some important hints to remember when using a floorplanning tool:

- Initially, all pads at the top level are assigned to be connected to a single layer. This constraint affects the router in its ability to optimize the connections to the pads. In some cases the pads can be assigned to two layers, which gives the router more freedom. In the case of many available layers (up to 10!), the assignment of pads to routing layers is best left to the layout designer and assigned manually. There may be dedicated layers for some ports that need to be assigned, such as power supplies and clocks.

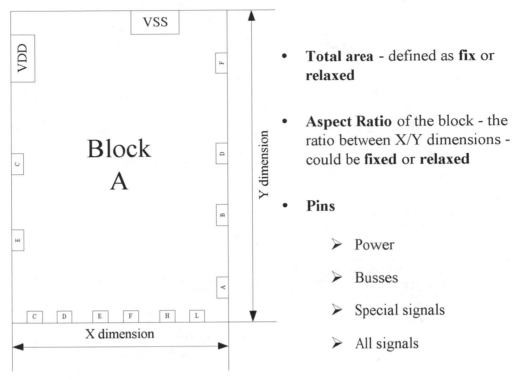

Figure 10.3 Block information needed for floorplan.

- Block pins may have to be manually assigned to the correct layer as in the previous case.
- Inserting spare logic and signals is tricky. Typically, the netlist of a design does not contain spare elements, so these elements need to be inserted manually as well. The person using the tool should learn how to "massage" the netlist by introducing "fictional" lines and pins, enlarging blocks for spare logic, and adding dummy blocks that will be used for spare logic. Proper placement of the spare elements is also necessary to ensure that the spare lines are close to the spare logic so they can be used easily.
- Blocks that are subject to change should be placed strategically where there might be extra space for the block to grow in size.

10.2.2 Block Floorplanning Tools

At the block level, the procedure for floorplanning is fundamentally the same as the chip floorplanning procedure, with minor exceptions. Instead of working with blocks, the floorplanner works with groups of cells that are defined based on the functionality and connectivity of the design. In this case the design is more mature, in that the design is better defined with cells and connections and there are fewer empty blocks with unknown contents.

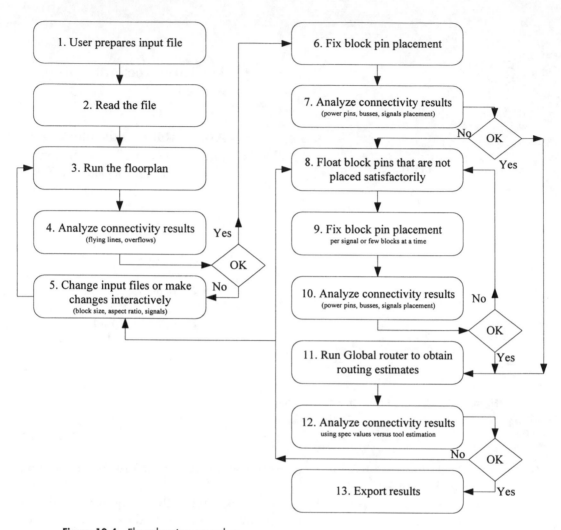

Figure 10.4 Floorplanning procedure.

In comparison to chip floorplanning, block-level floorplanning has the following similarities and differences:

- At the block level there may be thousands of cells; at the chip level the number is usually less than 20
- Port handling is similar, although there may be many more ports
- The pins in a block can number in the hundreds of thousands, and the connectivity is much more complex than at the chip level
- If the process has more than three layers of metal, the floorplan may not have channels at all—the plan will look like a sea of cells
- In the case where there are imported or hard cores, the floorplanner will optimize their placement among the rest of the logic.

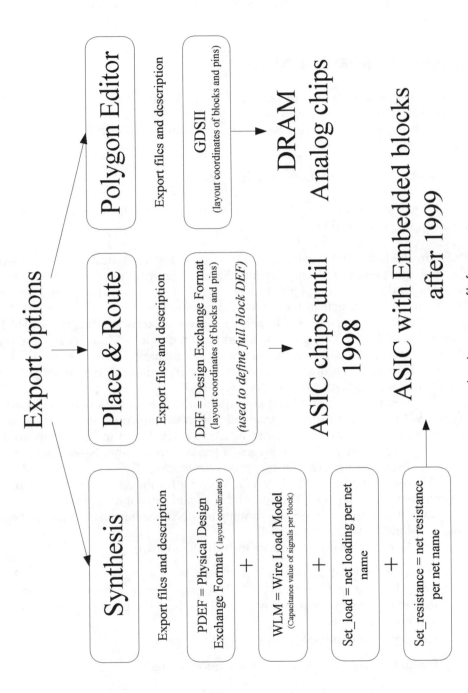

Figure 10.5 Floorplanner output file formats.

Floorplanners need to export information based on the design needs or the overall design flow being used for the design. Figure 10.5 illustrates the data formats that a floorplanner provides. The different formats depend on the target application.

10.3 LAYOUT GENERATION TOOLS

Layout generation tools are the main bulk of the layout design tools on the market today. Different tools are used depending on the level of abstraction, and this ranges from cell- or transistor-based layout to chip assembly. Each of these will be covered in this section.

10.3.1 Cell-Level Layout Generation Tools

Cell-level layout generation tools manipulate polygons to form transistors and connections between them.

Polygon Pusher. The most popular tool for cell-level layout generation is still the "polygon pusher" or basic "layout editor." Theoretically, a polygon pusher can be used to design a complete chip of any kind from transistor level and up. However, it is not practical for block or chip assembly.

Polygon pushers are used to generate low-level cells, ranging from 10 to 100 transistors, and for drawing all the layers required in a VLSI process. These tools are very mature, and generally minimum maintenance is required. However, some companies have invested a lot of time to "enhance" the basic tool with macros to generate layout that is done repeatedly.

Polygon pushers provide an environment for very high-quality full-custom layout, but they are completely manually operated and thus are slow. Enhancements to the environment are constantly being developed in an attempt to add automation to speed up certain stages of the design process: N-well generation, contact cells, transistor generators, ports macros, etc.

Almost all the big IC CAD vendors offer a layout editor, so it is important to evaluate them in terms of the availability of add-ons for automation. With only a few exceptions, most of them are running on UNIX machines, but they are beginning to migrate toward PCs as well.

Applications for layout editors include the following:

- Full-custom VLSI layout, where *size* is the most important factor. Pitch limited layout in DRAM, SRAM, PLA, pad output buffers, and input protection devices are only a few examples where the pure polygon pusher is still needed as it is.
- Analog design and special process cells or keys.

Features of layout editors include the following:

- They enable the user to do layout design of any kind.
- They are easy to set up for any process and most of the time work out of the box.

- Minimal training is required to get started. Advanced layout designers designing analog circuits, for example, need to understand layout concepts more than features of the tool.
- Over time, most companies have customized and have built entire methodologies around their layout editors. This is the main reason why stand-alone automated tools have not been able to displace them so far.
- Layout generation is *very* process dependent, so it made sense in the past to do everything in a 100 percent full-custom manner. ASIC flows and place-and-route changed the way people thought in terms of layout design.
- There are many internal tools inside big companies for this specific task that require a CAD group to develop and maintain the design environment. In many cases political interests are against progress and efficiency. Startup companies do not have to deal with such problems, so they are easily accepting new tools.
- These tools are slow compared to automated tools, and that is why they are rarely used alone. More and more they are mixed with complex device generators, transistor place-and-route, symbolic editors, compactors, and routers. All the vendors who have a large customer base are working hard to improve them by adding some type of automation.

Device Generators.   Device generators with routers can be used to implement almost all levels of layout complexity. Most of them are very specific to low-level or transistor-level layout design in different applications.

The tool is used to generate low-level cells, and/or complete standard cell libraries, very fast and efficiently. In this case each cell is limited to ~10 to 100 transistors per cell and all the layers required in a VLSI process are drawn.

Setting up and maintaining this type of tool requires extensive knowledge. Device generators provide the best mix of layout automation and full-custom design possible for cell generation.

Advanced device generators available today are schematic or netlist driven so the result is correct by construction. Knowledgeable drivers (layout designers) can work wonders using this "enhanced layout editor."

Applications for device generators include the following:

- Standard cell libraries, where standardization of the pin assignment, cell height, neighboring requirements, etc., is an important factor in layout design
- Cells for datapath, where the tool and design requirements have to be guaranteed and tailored to specific designs
- Applications where quality and analog requirements are as important as layout design speed

Features of device generators include the following:

- Layout generation is fast, but the tools are expensive for a small company. As we write, the prices of these type of tools are dropping because of competition. This is supposed to be the next generation of the basic layout editor.

- The tools have to be set up by an experienced user, but then can be used by any new trainee. For special layout requirements, experts in layout design and macros may be required, but these cases are typically only 10 percent of the effort in cell design.
- They are fast compared to full-custom polygon pushers, which is why they have gained market share in the past 5 years; however, they are slower than silicon compilers.
- These tools provide a good environment for process migration and/or process changes during the design process—processes evolve during the 1 to 2 years of a project's design time.

In principle, the market for silicon compilers is moving toward regular structures and block-level layout, while device generators are aiming to replace the old polygon pusher.

Cell Placers. Cell placers are used for optimizing the placement of individual devices and layers. In many cases a device generator is combined with a cell placer and a cell router to provide a complete layout environment. Ideally, cell placers need to understand cell architecture for good results.

Following are some features of cell placers:

- The numbers of devices and pins that these tools handle are small (hundreds), but the placement optimization is very detailed. Many constraints are considered.
- Ports can be placed on all sides. The placement may be fixed or may "float," in that it can be changed if a better layout can be achieved.
- Regions for PMOS and NMOS transistor placement are controlled—for example, transistors can be placed in rows of variable heights and may include multiple rows of different heights.
- Transistor fingering is automatic, based on specified architectural constraints.
- Substrate and well connections can be controlled.
- Layout that is already complete can be imported. This is useful in the case where a rough layout is done by hand and given to an automated tool to finish the job.
- The interface to the cell router is seamless.

At this level, placers really enhance the layout designer's speed and productivity. There are very few noticeable disadvantages of using placers in cell-level layout. Using a placer is a big step forward in increasing the amount of cell-level layout automation. The only barrier to their use is that they have to be easy to set up and use, and because they are used by designers who are very comfortable with polygon editors, these tools need a very good graphical user interface. Historically, all text-based tools have failed here.

Cell Routers. Routers for cell-level layout are typically very simple. However, some vendors have carved a niche by promoting the features of cell-level layout

to the requirements of chip-level assembly. There are a few good cell routers on the market, and it is important to have a device generator, a placer, and a router in the same environment. If driven by a netlist, cell routers can accelerate cell generation. The most powerful routers allow the user to define constraints such as the following:

- Equal length of signals—a full bus of signals will have exactly the same lengths for all bits
- Special line widths for predefined signals or applying special requirements interactively
- Power routing constraints
- The number of contacts for each connection and/or the minimum contacts for a source/drain
- Differential pair routing

Compactors.   A compactor can be used at almost all levels of layout complexity. Most of them are best used at a transistor or cell level. The tool is used to compact transistor layout and their connections inside a cell design.

One approach to using this type of tool is for the layout designer to do a loose job and run the compactor to optimize the layout. This is a very fast and efficient methodology to generate DRC clean layout cells.

For cell-level layout, the setup and maintenance of a compactor requires a very knowledgeable designer. The advanced compactors that are available today, together with schematic or netlist-driven layout generators, can provide the best of all worlds because the result is correct by construction and should pass both DRC and LVS checks. In the case where the compactor works on symbolic layout data, the results are extremely fast, and they can add advanced structures such as jogs within a wire if required.

Compactors are used at any level of design—from transistor-level layout to top-level routing, compactors should be part of any layout environment! Features of compactors include the following:

- They are capable of generating very quickly most kinds of layout, as well as correcting DRC errors before the errors are made.
- Any novice can use the tool if experts have properly validated the setup of the tool.
- Compactors are sometimes used for a limited process migration and/or process changes during the design process. When processes evolve during the 1 to 2 years of a project's design time, a compactor can fix minor design rule changes.
- For the moment, compactors are running in flat mode or within only one level of hierarchy. There may come a day when a "push-button" tool will be able to compact a full chip to fix a design rule change.

Silicon Compilers.   Silicon compilers can also be used to design cells of various levels of layout complexity. Similar to compactors, most compilers work best at the transistor or cell level of layout design. The tool is used to generate low-level cells, and/or standard cells that are limited to ~10 to 40 transistors per cell, very quickly and efficiently while drawing all the layers required in a VLSI process.

Silicon compilers require extensive and expert maintenance to be effective in a changing environment. They provide the fastest cell generation possible. A standard cell library can be generated in ~1 day while customizing the cells for a specific placer and/or routing tool.

Applications for silicon compilers include the following:

- Standard cell libraries, where standardization of the pin assignment, cell height, abutment, etc., is an important factor in layout design
- Cells for datapath, where the tool and design requirements have to be guaranteed and tailored to specific designs
- Any time speed is the most important factor in a layout generation

Following are some features of silicon compilers:

- Layout generation is fast, but the tools are expensive for a small company. For example, silicon compilers for standard cell libraries are so expensive that only companies selling libraries as their main product can justify their purchase.
- Only highly trained people in software, i.e., software engineers and/or designers with broad background in software, can use the tools. They may know how to run the tool, but not necessarily how the layout is supposed to look and be used. We hope this book will help them to understand more about generating layout for the entire design process.
- There are many internal tools inside big companies for this specific task that require a CAD group to set up, develop, and maintain the design environment. Standard formats are used to interface the output of silicon compilers with the other tools in the flow. The problem with using standard formats is that specific information that is required for the compiler may be lost and the advantage of using the tools defeated.
- They are so fast compared to full-custom polygon pushers that they have gained a lot of market-share in the past 5 years. Silicon compilers can be used not only for layout generation, but for process porting as well.
- Some compilers are targeted to specific applications: RAMs, ROMs, PLAs, I/O cells, standard cells, datapath designs, etc. These compilers do not require as much training because they have been designed for novice users.

Consider silicon compilers to be a suite of tools that contain device generators, placers, and routers under the hood. They do not offer the capability of interactive editing that is available by using the combination of the three individual tools.

10.3.2 Block-Level Layout Generation Tools

In general, block-level layout does not deal with transistors but with small cells and macros that are built with one of the tools described in Section 10.3.1. At the block level of layout design there are three types of tools. For regular structures, tilers are described, and for nonregular placement and connectivity, placers and routing tools are described.

Tilers.   A tiler is used to automate the generation of very organized and repetitive structures such as memories, datapath circuitry, and pad frames. The GUI is generally quite primitive, and they are available as a stand-alone tool or integrated in various design tools environments.

A tiler is a simple placer that understands a predefined or preprogrammed architecture and places leaf cells in the arrangement that is defined. A tiler simply executes a set of layout instructions (tiling program) and does not optimize or reconfigure a design based on a netlist or constraints. The layout generation part of a memory compiler is the most common form of a tiler.

Advanced tilers support decision constructs (i.e., if–then–else) and parameters in its tiling language. This feature enables a tiler to be configurable based on a set of parameters, and this is the basis for flexible memory compilers. For example, if the choice of cell is dependent on a specific parameter, then this can be built into the tiling program. The strength of a buffer can be chosen and tiled appropriately this way.

Following are some applications for tilers:

- RAMS, ROMS, PLA, I/Os, DATAPATH, etc.
- New migration tools are using tilers to break apart a particular design hierarchy and provide this information to the migration tool so that the leaf cell abutment constraints can be automatically generated.

Features of tilers include the following:

- The power of a tiler can be used as a very fast area estimation tool. A representative set of cells can be tiled quickly to obtain an estimate of area for a chip or a block. If routing is added and extracted, an estimate of interconnect delays is possible and is especially valuable in an ASIC design flow.
- Any novice can use the tool if experts have properly validated the setup of the tool.
- Tilers are generally not expensive.

Block Placer.   At the block level, placement issues are quite a bit different from those at the cell level. The block placer is one of the main tools used in an ASIC design flow. They are commonly known as the placement part of a place-and-route tool and are critical to the future of layout design as chips grow in size and complexity.

The size of the netlist is the first difference and in this case may be hundreds of thousands of cells and signals with many thousands of ports entering the block. This size of design is impossible to do manually in a reasonable amount of time. In practical terms, manual methods are limited to blocks with ~500 cells.

The block placer is a sophisticated tool to be able to handle large designs. This tool uses complex algorithms to optimize and reoptimize the placement of cells to achieve a placement that will meet timing, area, and routability constraints. This is not a simple task, as the tool needs to manage these trade-offs effectively.

One way to understand the role of a placer is to understand the inputs to the tool:

- A netlist of the circuit design that contains a list of cells to be placed and the logical connectivity between the objects to be placed.
- A description of each cell in the design. The description includes characteristics such as size, pins, pin locations, power consumption, and timing characteristics.
- Total available area and placement of ports to the block. In the case of cell-based placement the placer needs row information such as size, direction, channel restrictions if any, and location of black boxes or hard macros.
- In the case of a gate array, the location of the legal sites where cells are allowed to be placed.
- If advanced features such as power grid evaluation, IR drop per row, or electromigration rules are to be used, the constraints of each cell and for the entire design.

The output of a block placer is a preliminary version of a design where all cells have been placed in a specific location and the design is ready for routing to be completed.

Placement based on timing constraints is available and is a new and much more stringent methodology to use. In this case the placer has to evaluate placement based on the timing constraints received from the design file.

Block placers typically have a useful graphic user interface and are generally easy to set up. Block-level placers are essential to increase the productivity in layout design.

As discussed in detail in the next section, routing approaches have been either channel- or area-based. Channels are routing areas between cells, while area-based routers use all available area to route a design.

The placer that is used in a particular design must work closely with the router and must understand the limitations of the router. For example, a placer optimized for channels should not feed a completed design to a maze router.

Channel-based placers work on the basis that an infinite amount of area is available for routing and that routing channels can be expanded or compacted to accommodate and optimize the size of the channels. Cells are placed in rows, and the row that is chosen for a specific cell is determined based on the best place for routability and other constraints such as timing or power. The placement algorithm can vary the number of rows and the length of the rows to achieve a design that will meet all of the routing constraints.

Maze-based placers simply place all cells in rows within a fixed area without regard to dedicated routing channels; they assume the routing is completely over the cells. The placement algorithm considers issues such as routability, congestion, and timing, among many other things.

Block Routers.   As the name implies, a router automatically completes the connectivity of a placed design. Connections are implemented between cells and the interface to the block or chip. Routers in general can be used at all levels of layout design for all methodologies.

Features of block routers include the following:

- Routers automate the task of connecting millions of signals while optimizing for things such as area, timing, and power. This capability cannot be replaced by manual techniques.
- Effectively using both routers and placers requires a significant amount of experience to take full advantage of the many features built into the routers available today. Routers are not effective right "out of the box" and most companies have dedicated experts for this task.
- Some designers are reluctant to give up control of the routing to an automatic tool. However, confidence in extraction methodologies is usually sufficient to alleviate concerns and the speed of routing is hard to beat. Routing techniques are useful for nontraditional applications such as analog, RF, and memories, and it may be market forces that promote the use of routers in these areas.

A brief history of routers is presented next. It is useful to understand the evolution of the tools because it can give us a lot of insight into the concepts behind routing tools today.

Historically, routers were initially developed as a tool to assist or automate existing block layout methodologies for processes that were available at the time. Only one or two layers of metal were available; thus, routers had to work within the same constraints as the layout designer.

Routing channels were used extensively. The first routers had algorithms specifically targeting channel routing and are now known as *channel* routers. These routers essentially optimized the routing in one direction, that being the height of the channel.

As more routing layers became available, different algorithms were developed to take advantage of the extra layers. With three or four available layers, channels became unnecessary because routing could be implemented directly over the cells and the cells could be placed adjacent to each other without wasted space in between. In this case, *maze* routers became dominant because the "maze" algorithms built into the routers optimized the routing based on a mesh or two-dimensional maze of routing resources.

Finally, more recently, *shape-based* routers have appeared to address the chip-assembly or transistor-level designs. Shape-based routers are much more powerful in terms of implementing customized routing, but they are very limited in capacity when compared to maze and channel routers. Channel and maze routers are able to handle large databases because they essentially work on a point-to-point basis and use a well-defined and coarse grid. Shape-based routers implement polygons, and therefore need to manipulate much more information. A description of shaped-based routers follows in Section 10.3.3.

In terms of block routing, channel and maze routers are most relevant to our discussion.

In Chapter 5 we described how the architecture of the standard cells defined how the routing was done, and we gave a specific example of layout using channel routing. How does a channel router work in this environment?

At the beginning, with only one or two routing layers, the first routers connected signals only between the boundaries of the cells. Routing was optimized within the channels to reduce the number of signals within any particular channel. When three routing layers became available, channel routers were enhanced to be able to route over the cells and take advantage of the fact that the ports of the cells were in the middle of the row.

Channel routers optimize the size of different channels as an initial analysis step and require a specific algorithm to do a good job. This analysis ensures that the routing can always be completed, but the problem with this assumption is that the block may end up being too large for the space allocated to it.

Channel routers then automate the layout of channels by understanding the routing architecture and its limitations. The tools offer additional features such as the following:

- Automatic addition of jogs in wires to reduce the total size of the block.
- Automatic reduction of the number of vias by optimizing any jumpers to preferred routing layers. This is a very useful feature when the routing is between two blocks with only busses between them and no other passing signals. This feature also helps reduce chip size and signal resistance.
- Adding via topologies based on a specified formula and not limited to a simple fill algorithm.
- For symbolic channel routers, easy manual switching of a signal from one channel to another and to run compaction to fix DRC errors.
- Timing-driven placement and power analysis (offered by most today).
- For some routers, built-in routines that interactively adjust power connections, tapering, routing widths and positions, etc. For small geometries, delay and power calculation is necessary.
- For other routers, timing analysis built into the tool that can be done before or after routing so the designer can adjust certain routing parameters before completing the final route. In general, the timing is based on pin-to-pin delays.
- Current calculations built into some tools that provide many sets of data such as absolute current in a wire, the current density in the wire, and supply voltage for every node and cell.
- In most routers, ECO capabilities. However, remember that a complicated change many increase the size of the block/chip!

Channel routers have some disadvantages as well. The biggest disadvantage of channel routers is the uncertainty of block or chip size, which is not fixed until all the routing is done. The second is that they impose restrictions on the design of the standard cell library. For instance, cell pins may be required to be aligned in a single row or to be positioned on the border of the cell.

As mentioned previously, channel routers became less effective when chips began to be manufactured using four layers of metals. At this time maze routers achieved maturity, and extraction in an ASIC flow became standard procedure. Using a maze router, a design can be cell limited and routing channels are not required. This feature made chip size estimation easier and the size of a block or design could be determined earlier in the overall process.

Maze routers do not work with channels, but attempt to complete the routing using an area-based approach where connectivity is optimized based on horizontal and vertical routing resources. The available area must be predetermined and constrained for the router so that signal lines stay within a fixed area.

Different algorithms are used within maze routers. Generally, a global routing algorithm is used to subdivide the total area, a detailed router to complete signal routing, and a clock tree routing algorithm and a power routing algorithm for power supplies. All of the algorithms are linked together to complete a final design.

Unlike a channel router, where a route will complete every time, there is no guarantee that the maze router will be able to route all nets. Routing congestion and/or impossible timing constraints are the likely culprits that prevent the completion of a routing job. The best way to address these issues is to modify the placement of the cells. In some cases, for the last 10 or 100 nets, human intervention may be able to finish the job.

Routability problems are best solved by changing the placement of the cells, and this is why a strong link between the router and placer is needed. In this case the placer can place objects that are router friendly, and this is only possible if the placer understands the algorithms of the router. This is why it is highly recommended that the placer and router come from the same company.

ECO functionality is available within maze routers to implement minor changes in the design. Existing nets can be locally removed and then reconnected based on the new design.

In terms of worldwide use, most ASIC houses are using maze routers not only for standard cells and gate arrays, but also for FPGA designs.

After the blocks are routed and simulated with extracted parasitics, it is usually time to insert any clock trees. The placement is generally the starting point for clock tree insertion, as the distribution of loads is known. In this case the routing tool has another algorithm to find the cells that are connected to a specific clock and is placing and routing only the buffers and clock signals to minimize clock skew. (See Figure 10.6.)

A measure of quality for place-and-route is what is known as *utilization*. The utilization of a design is defined by the ratio of the area consumed by cells to the total available area. Channel-based designs will have a lower utilization number because of the overhead of channels. Gate array designs limit cell density by having limited cell placement sites and in this case utilization factors will be in the range of 50 to 70 percent. Maze routers will have the highest utilization factor because of the flexible cell placement and over-the-cell routing. Nevertheless, the cells are placed in rows, and not every row will be 100 percent full, so the utilization will not be 100 percent.

Remember that placers that work with maze routers use a fixed die size and do not include a compaction step. Therefore, there is no way to improve efficiency of a design after all the routing is done.

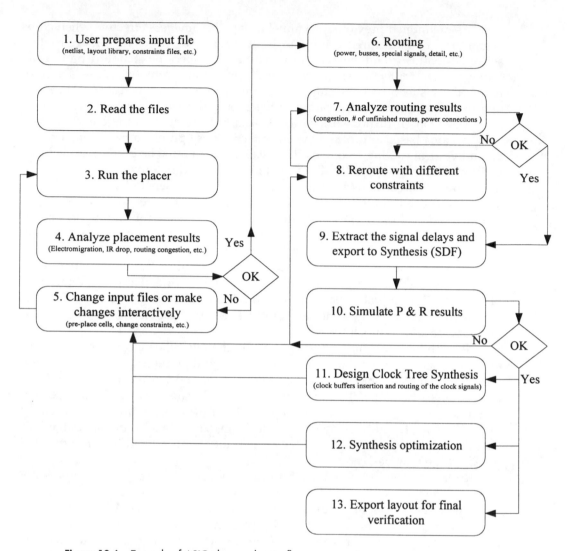

Figure 10.6 Example of ASIC place-and-route flow.

Combining the approach used in channel routers of routing analysis, placement then compaction with the efficient routing algorithms of maze routers provides the best of both approaches.

The concept is that a channel-based placement is done that fully evaluates the design for optimal routability, and after placement the "channels" or rows are compacted to determine the final block size. It is this final block size that the maze router uses to complete the routing. Since a channel-based placer has the freedom to define the number and length of rows, the routability of the final design will be much higher than the output of a maze-based placer. This is because a maze-based placer is given the available area and cannot change the aspect ratio of the block to improve routability.

With this optimized design size, the maze router will more easily complete the routing and the overall effect should be an improved utilization factor. The big players (vendors) in the place-and-route tools market are trying to implement a similar approach in different ways.

10.3.3 Chip Assembly Tools

Chip assembly, as the name implies, is the process of combining the different blocks of a chip with the pad cells and integrating them all into a completed design. Consider the following scenario:

- The layout leader designed a chip floorplan using a tool for placement of blocks.
- The blocks had their pins identified compatible for routing.
- A router was used to connect the blocks.
- The plan was updated during the course of a project by doing the following:

  As each block was finished, its empty box was replaced with the final block layout

  The routing was updated to reflect the new position of pins and widths of the signals

An analysis of this process should conclude that the chip assembly was completed near the beginning of the project in the form of the "initial" floorplan! With the help of automated tools that are available today, this flow is not a dream, but completely possible.

In general, for top-level assembly or routing between blocks that number no more than 50 and pins and signals that number no more than 5,000, shape-based routers are the best tools for the job.

Shape-based routers are very rich in features, but have the disadvantage of limited capacity. Channel and maze routers simply connect pins and respect obstruction layers using simple paths on a coarse grid. Shape-based routers have many features that include connecting a signal of different widths using polygons. Therefore, shape-based routers need to process a lot more information, and this is what limits their capacity.

Shape-based routers produce layout that is full-custom. Historically, these routers were based on approaches taken from printed circuit board layout tools where very controlled and detailed layout is required.

Most routers are automatic, but more and more users are asking for interactive features for routing. Some shape-based routers are already providing features that have never been available before:

- One bit of one signal of a bus can be routed manually, and then the shape of this bit can be copied relatively easily to all the bits of the bus.
- Busses can be routed together as a group with options to run them at 45-degree angles or around corners to minimize bus skew or routing area.

- Routing of signals can be constrained by grouping them in classes such as busses, special groups, or clocks.
- User-defined routing rules for each layer as well as net-based routing constraints address analog signal crosstalk, minimum capacitance, and resistance effects.

These routers were the first to offer automatic shielding to reference signals and diode application against antenna rules, thus demonstrating their PCB heritage.

10.4 SUPPORT TOOLS

In terms of layout entry, we have covered the basic types of tools involved. The design process is not only layout entry. The layout needs to be verified against various quality standards, and manufacturing design rules and layout can be migrated from other sources to save time and effort. Tools that perform all of these operations will be covered in this section, as well as a discussion of standard database formats used for layout today.

10.4.1 Layout Verification Tools

Figure 10.7 documents the layout design procedure as discussed in Chapter 3. The highlighted steps correspond to a verification tool that is discussed in this section. It is important to remember that layout verification must be done on the entire chip and on the file that is to be sent to manufacturing.

DRC/LVS/ERC. The verification of a full chip database file has been an issue for the verification tools over the course of time as designs have grown from 1,000 to 10,000,000 transistors. As the designs have grown, designers and tool providers have evolved the methodologies capable of verifying these designs.

From a user's point of view, the requirements for verification tools are different from those for layout entry or design tools. Ease of debugging or correcting problems, tool capacity, and run times are the key issues for these tools. The verification process in general is a feedback mechanism for the designer to validate the design as well as identify problems or shortcomings. Historically, the layout verification tools had limitations in addressing the key issues as summarized in Table 10.1.

In all cases the accuracy of the checks depends on the values and algorithms that are coded into them. These values and algorithms are captured in files referred to as setup files, command files, or rule decks.

"Work" structures refer to the methodology of breaking a full chip layout into smaller structures for verification that together try to ensure that all potential problems are found. This methodology is very time consuming and potentially error prone, but was necessary because the capacity and speed of the verification tools could not handle an entire chip at once.

The user interface of these tools has evolved significantly over time and they are approaching the ideal conditions just listed. For example, DRC error bars are

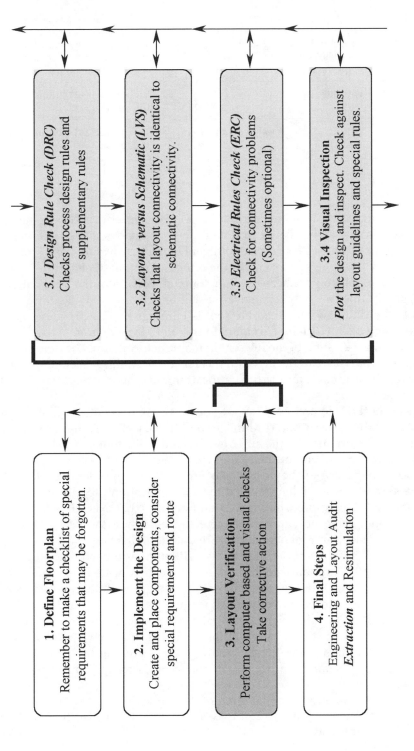

Figure 10.7 Layout verification tools.

TABLE 10.1   Verification Tool Issues

Key Issue	Ideal Methodology	Historical Methodology
User interface	Graphical debugging environment with visual links to circuit design and suggestions for correction	Very first tools generated text file output, but interactive debugging within polygon pusher is now standard
Capacity	Unlimited	Practical limit ~1 M transistors, so "work" structures were used
Run time	<1 hour so that repeated iterations are possible	Days or weeks depending on type of verification

standard and they show the location of polygons or edges that violate a particular rule. Cross-probing between layout and schematics is also standard for LVS debugging. Features of today's debuggers include browsing based on error type, layer, layer groupings, jumping to the critical errors first, etc.

Many polygon pushers have an "online" version of the verification tools so that small jobs can be executed almost interactively and the time required to export the database to a separate tool is eliminated. The capacity and run-time issues have been virtually eliminated for small blocks.

Final verification of the file sent to manufacturing must be done with a stand-alone tool that can read the tape-out file. Historically, this was not possible because the capacity and run times of the verification tools were not sufficient to check the entire design.

Recently, "hierarchical" layout verification has been introduced specifically to address the capacity and run-time issues. In the past the limits of the verification tools were determined by the amount of data that the tool had to load and process, and this was a function of the size of the design in terms of polygons.

The hierarchy of the design was ignored and any hierarchy that existed in the design was essentially removed for the verification process. A "flat" database was created. This approach ensured that polygons that existed on top of cells were checked alongside polygons that were drawn inside the cells. The tools were required to store and manage the entire database this way.

Hierarchical layout verification is a different approach that takes advantage of the hierarchy built into the design. Cells that are repeated are checked only once and then discarded for the remainder of the design. The tool requires careful management of the effects of over-the-cell routing.

Note that layout designers can take advantage of the features of the hierarchical verification tools by building efficient hierarchy into their designs. Specifically, limiting over-the-cell routing and matching the layout hierarchy to the schematic hierarchy are good methodologies to accelerate the layout verification process.

The key issues within a hierarchical layout verification environment are reexamined as shown in Table 10.2. It may appear that layout verification issues have been completely solved!

TABLE 10.2 Hierarchical Verification Tool Issues

Key Issue	Flat or Historical Methodology	Hierarchical Methodology
User interface	Every error is reported independent of repetition	Errors that occur in a repeated cell are reported only once
Capacity	Practical limit ~1M transistors, so "work" structures were used	No limit, especially if repetition and hierarchy are used. "Work" structures are not necessary. Verification of tape-out file possible
Run time	Days or weeks depending on type of verification	Hours in all cases

One place where LVS debuggers can improve is the location and debugging of power-to-power shorts. This type of error is very pervasive and produces a lot of output if left uncontrolled.

Extraction.   Extraction is the hottest product today in the Deep Sub-Micron era. Layout extraction is another way of verifying that the layout performs as expected. If DRC checks the rules for mask making, and LVS checks that the connectivity and sizes of all devices are correct, the extraction of the layout is checking that the performance of the layout in simulation meets the required goals.

Layout extraction produces data that feeds back the result of layout to the circuit design process. The format of the data can be simply a netlist of devices, resistors, and capacitors, or the extraction tool can simplify the network of parasitic components by calculating an equivalent delay or lumped RC model.

Extraction is nice to have for normal digital circuits but is a must for analog, RF, and microwave designs where each small capacitance can change circuit behavior.

Extraction methodologies and tools evolved much more quickly in parallel with the development of ASIC flows, since the level of automation in circuit design was increased, thus separating the designer from manually designing all aspects of the design. The extraction process gives the required feedback to the circuit design to evaluate the layout implementation of the circuit. In the case of the ASIC flow, extraction of the real layout from place-and-route is crucial to the size and timing of the design. The reasons are obvious:

- ASIC designers are not analog experts, as they concentrate on developing functionality
- The number of nets in a design is impossible for a human to digest
- In general, ASIC designers do not even see the layout

Back annotation is the term that describes the step of feeding layout information back to the circuit design. Final simulations should be run with the extracted values from layout. For a final extraction to be successful, there are a few minimum requirements:

TABLE 10.3 Extraction Tool Accuracy Levels

Level	Description	Run Time
1D	L and W of lines only are used in calculation	Minutes per design
2D	Limited near-body effects are considered	Hours per design
Quasi-3D	Table lookup of parasitics based on predefined patterns	Days per design
3D	3D field solvers are used to calculate parasitics	1 day per net

- The layout is DRC and LVS clean without errors or warnings
- The extraction environment is set up with accurate process information and tested on a small circuit as a sanity check
- Critical signals are extracted with a higher degree of accuracy

The circuit design team should understand the accuracy of the extraction so that they can account for the limitations of the tool when modeling and designing their circuitry. Extraction tools trade off accuracy for run time as shown in Table 10.3. The main difference in accuracy is how the extraction tool calculates the effects of near-body effects. In the example, 1D near-body effects (i.e., coupling to other lines) are not considered at all. 3D field solvers not only take into account all near bodies, but also solve complex sets of equations to calculate parasitic values.

Ideally, the extraction flow is very fast and perfectly automated so that seeing the layout becomes unnecessary. In this case the setup of the flow is very important.

Note that visual audits should still be done for many specialized applications. People's eyes and expertise are still useful to analyze the effects of different layout architectures. The only proven methodology today is plotting the cells/blocks and asking experts to audit the layout.

Plotting and Plotters.   There are not too many kinds of plotting software available in the market. In general, layout designers are using two kinds of plotting software. One is simple printing software that is using the drivers of the specific printer to print the layout. The more expensive but extensive version is software that is written for "plotting" VLSI. Let's see what are the advantages of such software:

- The user can plot parts or "windows" from the big cell
- The user can choose only specific layers to be plotted
- The user can define fill patterns different from the ones shown on the screen—in general, designers are using a black background for screen work and white for plotting, so what looks good on black may not on white, especially when there are 3 to 10 layers of metal on top of each other
- The user can define a scale for plotting such as 1,000×, 5,000×, 10,000× so the picture will be greatly enlarged to analyze analog problems

- The user can choose a variety of options for plotting cell arrays—for example, doughnut shapes or corners only

Why do we need plotting at all? One reason is that there may not be coding for all of the rules for DRC verification. The second is that in some cases some rules are very rare and the DRC rule check may not be deterministic. The third important reason is that architectural improvements may only be understood from a visual inspection of a large-scale plot.

In terms of plotters, there are two kinds available today for VLSI applications: electrostatic or inkjet. Electrostatic plotters require a climate-controlled room with a high level of humidity. This is one reason why inkjet plotters are becoming popular, as they work at room temperature. However each of these plotters has advantages and disadvantages:

- Electrostatic plotters can deliver perfect plots up to 10 meters in length.
- Inkjet plotters have a limit to the length of plot.
- Electrostatic plotters are more expensive to buy, but in terms of price per square meter of plotting area, the cost is the same in the long run because the ink and the paper are more expensive in the case of the inkjet.
- The widths of the plots that can be obtained are comparable because all of them have 36-, 44-, or 54-inch paper width capability.
- Electrostatic toner is delivered in gallon sizes compared to ink that comes in 1.36-liters bottles.
- Both types can be connected to the network and organized in a queue for plot prioritizing.

In conclusion, if you do not need a color plotter to perform audits, you may not need a plotter at all. Black-and-white plotters are used mostly for mask/reticle check, where they are even 64 inches wide, but there are only two layers to check against each other. When you have to deal with four layers of poly and two or three metals, as in a DRAM process, color is obligatory.

10.4.2 Migration Tools

Migration tools are most useful in three scenarios:

1. Second sourcing for added capacity or reliable supply
2. Design reuse
3. Manufacturing process evolution

Lately, silicon intellectual property has become "in" in the VLSI industry, and layout converters have really started to get global attention. There are two ways to deal with process migration. One is to design layout inside tools that can perform process retargeting; the other is to use GDSII converters after everything is silicon proven.

A converter can be used to change almost all levels of layout complexity. The tool was used in the past to migrate low-level cells, and/or full standard cell libraries, very fast and efficiently. Each cell was limited to ~10 to 40 transistors per cell, and all the layers required in a VLSI process were converted. For such tools, extensive and very knowledgeable setup and maintenance is required.

Converters provide the best solution to migrate full chips from one process to another. The converters available today are *not* schematic or netlist driven; however, some transistor resizing is possible using tables or scaling factors. In general, the cell topology, pin positions and assignments, electromigration, and RC delay requirements are maintained.

These tools are used for purposes such as the following:

- Standard cell library migration, where standardization of the pin assignment, cell height, neighboring requirements, etc., is an important factor in layout design.
- Cells for datapath, where tool and design requirements have to be guaranteed and tailored to specific designs.
- Full chip conversion. Converters are starting to work hierarchically, so the size of the source data is no longer a problem.
- In the case where the chip is in advanced stages of layout but a process design rule that affects chip size is changed. Running a converter in hierarchical mode will solve the problem in a matter of hours with almost 100 percent DRC correct results.

Advantage and disadvantages include the following:

- They give the user the capability to migrate specific kinds of layout quickly, but they are expensive for a company that works in a single defined and proven process. Startup fabless companies will likely invest in migration tools, and this will fund further tool development.
- The user may require a minimum amount of training in the macro language, but advanced layout and design knowledge is key. It is important to understand the key characteristics of the source layout to ensure that the target layout quality is maintained.
- This type of tool may take some time to set up and to interface with other tools involved in the design flow.
- They are fast compared to any other full hand-crafted capability in migrating layout, and that is why they have gained so much market share in the past 5 years.
- The drawback is that these tools cannot add layers—i.e., migration of a two-layer metal chip to a three-layer metal chip.
- They fully respect the original topology; however, they cannot take advantage (alone) of new and perhaps better architectures that may arise in the destination process. In the case of libraries, the easiest solution is to change the source with minimum effort and then to run the conversion. In case of full chips, the vendors of this kind of software developed various levels of migration, such as only the cells, cells and routing, and only routing.
- We should chose the tools based on capabilities, but also on the user interface and setup simplicity. If the migration is efficient but takes time to set up and to debug the constraints, then the total effort is what counts.

- Interestingly enough, some silicon compiler vendors who totally ignored the migration market started to work in providing GDSII input and output to and from their tools to grab a piece of this "hot" new pie.

10.4.3 Data Formats

Any designer who wants to use point tools instead of integrated tools within one framework has to learn how to deal with the data transfer issues. Every layout tool starts from a different idea and they all have a different purpose, so internally they each may have a database format that is efficient for their needs.

For example, a problem starts when a layout designer wants to transfer data of a standard cell library to a different place-and-route tool. At the beginning of the IC design industry, there was only one company providing layout design tools for the entire market. This market was very small compared with today, and the format was defined by them based on the limitations of the hardware and software of the time. Everybody who wanted to enter the VLSI layout market had to comply with this format; otherwise, no one would buy their tool.

The format was and still is GDSII and was developed by Calma on Data General machines. So today if you want to export data from the Mentor platform to the Cadence platform, the only guaranteed way is GDSII. There are other widely used standard data formats such as CIF, LEF, and DEF, but they became popular for the same reason: Cadence developed these formats and had the greatest market share for IC layout. The GDSII format is still the dominant format, so a discussion of this format is warranted.

GDSII is a binary format that, from the user point of view, has the following qualities:

- In each stream file there is a limitation of 64 layers, which have a subdivision of 64 DATATYPEs per layer. So in total the limitation of the stream is $64 \times 64 = 4,096$ different layers to define polygons for manufacturing.

- Each polygon or path cannot have more than 199 vertices, so if the layout has polygon bigger than this number, the output subroutine will break it into pieces of 199 only. This limitation comes from the Calma software, which could handle only 199 coordinates per polygon!

- There is no logical or electrical information attached to a polygon. There are no pins, ports, nets, or signal recognition, and this is a big drawback for place-and-route. There are no pins; however, there is a simplified form of recognizing ports. When writing a GDSII file the ports become TEXT with a TEXTTYPE that is attached to a small polygon on the layer specified in the export command file. When importing this GDSII into another tool, the user usually writes macros that will select the text and regenerate the ports. The problem is that there is no solution for net information to be preserved.

- Device generator results, vias and contacts, or automated layouts that are "not polygon level" or soft devices are flattened to polygons. This is a big problem when the transistor size is changing and the data is coming out of a tool that uses these features. Again, this is because historically, Calma didn't have device generation capabilities.

- GDSII recognizes full hierarchy of objects, but always takes the first reference cell found in the design, regardless of the full path. GDSII uses unique names for each cell but does not recognize the full path name, which is again historical. Unique names in UNIX mean that something in the full path name is different. In Calma times the cell names were attached to a library that had a unique place to be written on the disk.

Another format that is mostly used for place-and-route is LEF, which contains layout information required for a library and routing setups and, together with a DEF file, fully characterizes nets, pins, ports, and signals.

APPENDIX A

Audit Checklists

In the past 10 years, the tools for IC design have advanced tremendously and are trying to address all the new trends in processes, design flow, and methodologies. There are still a few areas in which the design and process requirements are not addressed using state-of-the-art CAD tools.

To make up for these discrepancies, there is still need for a visual review or technical audit. Over time, the list of items to audit has shrunk because the CAD tools have become more sophisticated. Because of the complexity of the audit task, only experienced designers are able to do this job effectively.

The secret to performing an effective audit is to use a specific and detailed checklist and to have as an auditor a person who understands the extent of the problem and can propose solutions. In general, the auditor should be a person who has not participated directly in the project so that any bias to design styles and methodologies is avoided.

In this addendum, we will try to help new or experienced designers by providing a checklist related to each level of complexity in the layout of a chip. Using a plot of the layout in question along with all relevant documentation of the layout, the auditor evaluates the quality of the layout and uses the checklist as a guide.

A general checklist that applies to all layout is provided in Table A.1.

A.1 CELLS

The audit checklist for cell layout includes items related to transistor layout issues and the design of the cell for use in a block-level design.

Table A.2 shows a generic checklist for a transistor-level layout design.

TABLE A.1 General Layout Checklist

#	Question	Answer
1.	Is the cell DRC correct?	❏ Yes ❏ No ❏ N/A
2.	Are there any DRC rules that should be checked by eye?	❏ Yes ❏ No ❏ N/A
3.	Is the cell LVS correct?	❏ Yes ❏ No ❏ N/A
4.	Are there any LVS rules that should be checked by eye?	❏ Yes ❏ No ❏ N/A
5.	Is the cell ERC correct?	❏ Yes ❏ No ❏ N/A
6.	Are there any ERC rules that should be checked by eye?	❏ Yes ❏ No ❏ N/A
7.	Are there any special requirements for the layout?	❏ Yes ❏ No ❏ N/A
8.	Is the critical path of the schematic respected?	❏ Yes ❏ No ❏ N/A
9.	Were layout guidelines followed where possible?	❏ Yes ❏ No ❏ N/A
10.	Were electromigration rules satisfied?	❏ Yes ❏ No ❏ N/A

TABLE A.2 Cell Layout Checklist

#	Question	Answer
1.	Is the cell designed to minimum dimensions?	❏ Yes ❏ No ❏ N/A
2.	Does the cell follow a standard template?	❏ Yes ❏ No ❏ N/A
3.	Are the power lines notched anywhere in the cell?	❏ Yes ❏ No ❏ N/A
4.	Are all poly lines as short as possible?	❏ Yes ❏ No ❏ N/A
5.	Do transistor source/drain areas have enough contacts?	❏ Yes ❏ No ❏ N/A
6.	Is the transistor fingering optimal?	❏ Yes ❏ No ❏ N/A
7.	Are there sufficient substrate and tub contacts?	❏ Yes ❏ No ❏ N/A
8.	Are there any soft-connected nodes?	❏ Yes ❏ No ❏ N/A
9.	Are all the ports properly assigned by project standards?	❏ Yes ❏ No ❏ N/A
10.	Is the cells interface designed to ensure proper connectivity?	❏ Yes ❏ No ❏ N/A
11.	Is the origin in the lower left corner?	❏ Yes ❏ No ❏ N/A

A.2 BLOCKS

The block layout checklist addresses more global problems related to the block type and connectivity between cells. Different blocks will have different requirements to check for. For example, full-custom blocks, blocks of standard gates or cells, datapath, register file, or multiplier blocks will have specific checks related

TABLE A.3 Block Layout Checklist

#	Question	Answer
1.	Does the block follow the floorplan?	❏ Yes ❏ No ❏ N/A
2.	Was the power grid defined from the result of a simulation for power consumption, electromigration, and RC requirements?	❏ Yes ❏ No ❏ N/A
3.	Is the power supply strapping adequate and implemented with enough vias?	❏ Yes ❏ No ❏ N/A
4.	Are the power lines notched anywhere in the block?	❏ Yes ❏ No ❏ N/A
5.	Is the length of all critical path signals optimized?	❏ Yes ❏ No ❏ N/A
6.	Have all special signal requirements been satisfied?	❏ Yes ❏ No ❏ N/A
7.	Are all the ports properly assigned by project standards?	❏ Yes ❏ No ❏ N/A
8.	Is the block interface designed to ensure proper connectivity?	❏ Yes ❏ No ❏ N/A
9.	Are there adequate spare lines and logic?	❏ Yes ❏ No ❏ N/A
10.	Are there probe pads for specified signals?	❏ Yes ❏ No ❏ N/A
11.	Is the origin in the lower left corner?	❏ Yes ❏ No ❏ N/A

TABLE A.4 Chip Layout Checklist

#	Question	Answer
1.	Does the chip meet all packaging requirements?	❏ Yes ❏ No ❏ N/A
2.	Are the power supply connections to the pads adequate?	❏ Yes ❏ No ❏ N/A
3.	Is the power supply strapping adequate and implemented with enough vias?	❏ Yes ❏ No ❏ N/A
4.	Are the power lines notched anywhere in the top-level routing?	❏ Yes ❏ No ❏ N/A
5.	Is the length of all critical path signals optimized?	❏ Yes ❏ No ❏ N/A
6.	Have all special signal requirements been satisfied?	❏ Yes ❏ No ❏ N/A
7.	Are there adequate spare lines and logic?	❏ Yes ❏ No ❏ N/A
8.	Are there probe pads for specified signals?	❏ Yes ❏ No ❏ N/A
9.	Is there any sensitive circuitry placed close to the edge of the die?	❏ Yes ❏ No ❏ N/A
10.	Are the rules for the chip corner areas satisfied?	❏ Yes ❏ No ❏ N/A
11.	Is the interface of the chip to the scribe line properly defined?	❏ Yes ❏ No ❏ N/A
12.	Have all ESD and pad latch-up requirements been satisfied?	❏ Yes ❏ No ❏ N/A
13.	Have all the necessary chip finishing cells been included?	❏ Yes ❏ No ❏ N/A
14.	Is the origin in the center of the die?	❏ Yes ❏ No ❏ N/A

to the function of the block. The block layout checklist shown in Table A.3 outlines questions for general issues that are important in most cases.

A.3 CHIPS

For a layout audit at the chip level, the level of complexity of the audit is even greater. In this case there is a great variety of issues to verify that depend on the type of design. For example, the list may be shorter for an ASIC and very long for a full-custom analog chip. The various processes and methodologies are so different that it will be impossible to cover all of them. The chip layout checklist shown in Table A.4 outlines questions for general issues that are important in most cases.

APPENDIX B

Database Management

Throughout the course of a project, there are teams of people working together and in parallel on many different aspects of the design. Many different kinds of data are created, revised, shared, and deleted very dynamically and quickly. The data can be layout data, but also includes schematics, setup files, documentation, and many other kinds. How is all of this data managed? Formal database management techniques are the answer.

Database management is a process supported by an infrastructure that fundamentally provides the following features and benefits:

- *Version control:* Each file that is managed should have a version associated with it.
- *Version history:* Histories of all database objects are tracked.
- *Data sharing:* Data must be shared, and this should be done in a systematic way. Changes to shared cells must not affect work in progress that uses the cells.
- *Database integrity:* Protection against inadvertent deletion or database corruption. Examples of database corruption would be missing cells referenced in a hierarchical design or two cells of the same name in two different places on the computer system.

It is the last point that is the main reason for a formal database management process. As blocks are finished and the tape-out date nears, database integrity is crucial.

Almost every company has a different way of dealing with this problem. However, the issues and concepts are the same as those just listed. A basic approach to database management is presented, as well as, for ease of understanding, the scenario of layout database management.

In a practical sense, in IC design there are a few basic concepts that make the system work. The function of the database management is to ensure the following:

- In general, there is a one-to-one correspondence between a circuit and layout design.
- Shared or lower level cells are frozen before being used by other team members.
- Only one person is allowed to change a cell at any given time.
- During the time a cell is being changed, other team members can use an older version.
- New versions of objects are announced and communicated as required.

Proper database management relies on a computer system that recognizes groups of users so that authentication of the person accessing data is possible. The access rights of the data that exists on the system should be specified for three groups of users, depending on the type and application of the data:

- *World access*: Everyone—necessary for global data such as CAD software
- *Group access*: Team members—limits and identifies data specific to one project
- *Owner access*: An individual person who last modified or manipulated the data

The access rights for manipulation of the data should also be identified:

- *Write access*: Changes can be made to the data
- *Read access*: The data can be used or referenced but not changed

An example scenario is shown in Figure B.1 for the creation of a cell.

In this scenario Fred is a member of the layout team for the DSP32 project. Fred's job is to create a cell call AGBC. The key things to note are these:

- Fred checks to make sure the cell name is not already taken in the database. This is important because perhaps the cell layout has inadvertently been assigned to someone else or was already done. Fred can avoid some work in this case.
- The cell is not "checked in" until it is fully verified. Only at this point is read access given to the group. "Check in" is a term for releasing a cell for use by others, similar to returning a book to a library.
- Fred cannot modify the cell without "checking it out" from the database. Again, "check out" is the term for taking control of the cell from the database.

In this way cells are created systematically and the histories of cells can easily be maintained. Also, in the check-in phase automatic checks to the database can be built into the system.

The next scenario is more interesting, because now changes to cell AGBC are required (Figure B.2).

In this scenario Dan, Julia, and Brenda are also members of the layout team for the DSP32 project. Fred's job now is to change the cell called AGBC. The key things to note are these:

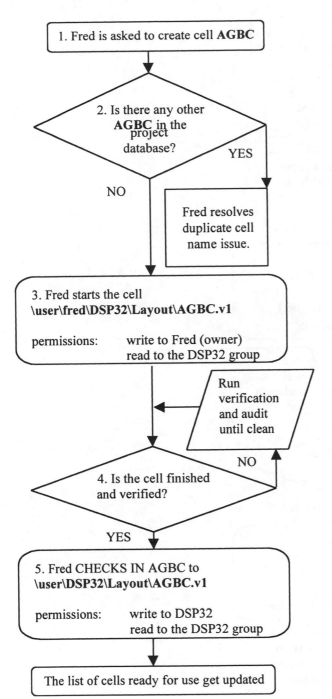

Figure B.1 Example steps of cell creation.

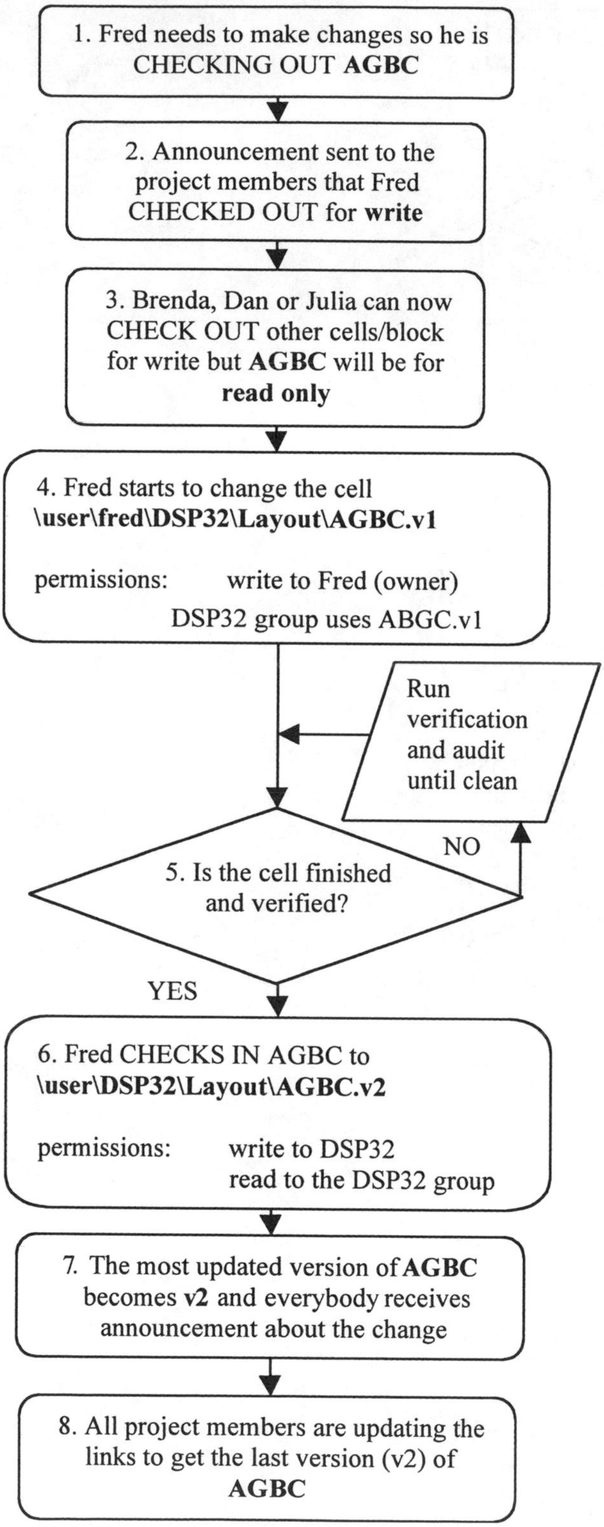

1. Fred needs to make changes so he is CHECKING OUT **AGBC**

2. Announcement sent to the project members that Fred CHECKED OUT for **write**

3. Brenda, Dan or Julia can now CHECK OUT other cells/block for write but **AGBC** will be for **read only**

4. Fred starts to change the cell **\user\fred\DSP32\Layout\AGBC.v1**

permissions: write to Fred (owner)

DSP32 group uses ABGC.v1

Run verification and audit until clean

5. Is the cell finished and verified?

NO

YES

6. Fred CHECKS IN AGBC to **\user\DSP32\Layout\AGBC.v2**

permissions: write to DSP32
read to the DSP32 group

7. The most updated version of **AGBC** becomes **v2** and everybody receives announcement about the change

8. All project members are updating the links to get the last version (v2) of **AGBC**

Figure B.2 Example steps of cell modification.

- Dan, Julia, and Brenda are notified that changes are forthcoming.
- Dan, Julia, and Brenda reference the old version of AGBC until the changes are complete.
- Dan, Julia, and Brenda are notified when the new version of AGBC is available and when appropriate can update their work.

The process of check-in and check-out to and from a central database will help all the designers in the project to keep working on their own assignments without being slowed down by various changes that are performed by others.

The systematic management of database changes is where these techniques are essential.

APPENDIX C

Scheduling

Scheduling the layout of a cell, block, or chip is in general a matter of experience. A few hints about scheduling layout tasks for your own work, as well as an idea of average industry speeds, will be presented in this section.

Forecasting a project schedule is a complex task that depends on many factors: tools, flows, hardware, design team experience, training, resources, holidays, sick leave, process changes, etc. In our experience the best way to accelerate any schedule is not faster computers or better software (they help), but increased expertise of the design team. There is no substitute for expertise for maximum productivity.

Let's think about scheduling the elapsed time for the layout of a cell. In all cases of scheduling it is important to think about the speed of an *average* designer and not the best performer, because the schedule has to reflect the reality of a varied design team.

For a cell-level scheduling, we need the following information:

- Number of transistors
- Number of signals
- Cell type—does it have a template to work from, or is it brand new?
- Special requirements

Now we can try to put numbers behind the requirements (Table C.1).

Note that the type of cell and the special requirements really affect the relative time it takes to complete the cell. Verification time is something that is not included in the tables because it is assumed to be the same for all scenarios.

In the case of blocks, there are different considerations:

- Number of components—cells, small blocks, random gates
- Number of busses, signals, and power grid requirements
- Special requirements—symmetry, crosstalk, minimum RC, timing

TABLE C.1 Example Cell Level Schedule in Hours

Cell Name	# Transistors	# Signals	Type	Special Req.	TIME Hand craft	Automated*
INV1	2	2	STD	None	4	0.25
INV × 20	2	2	STD	Min. capacitance	8	0.25
DFF	16	6	STD	Min. size	16	1
INV1	2	2	None	None	1	0.25
INV × 20	2	2	None	None	2	0.25
DFF	16	6	None	None	8	0.5
INV1	2	2	Datapath	Min. size	4 to 8	0.5
INV × 20	2	2	Datapath	Min. size	4 to 8	1
DFF	16	6	Datapath	Min. size	8 to 24	1

*Add to the project schedule the tool and architecture setup time that does not exist in the handcrafted schedule for cells.
TIME = Direct hours not including overhead (meetings, breaks, lunch)
STD = Standard cell for a library
Datapath = Register files, multiplier where there are many N/P/P/N regions
None = Normal random logic
NOTE: This timing does not include verification time.

TABLE C.2 Example Block-Level Schedule in Hours

Block Name	# Cells	# Signals	Type	Special Req.	TIME Hand craft	Automated*
Controller	200	150	STD	None	40	2
Synchronizer	400	300	STD	None	80	4
Register file	10 × 32	8 × 32 + 30	STD	None	80	4
Controller	200	150	STD	Timing	60	4
Synchronizer	400	300	STD	Timing	100	8
Register file	10 × 32	8 × 32 + 30	STD	Timing	90	8
Controller	200	150	STD	Timing + size	70	5
Synchronizer	400	300	STD	Timing + size	120	9
Register file	10 × 32	8 × 32 + 30	STD	Timing + size	90	9

*Add to the project schedule the tool and architecture setup time that does not exist in the handcrafted schedule for cells.
TIME = Direct hours not including overhead (meetings, breaks, lunch)
STD = Using standard generated gates (standard cells type)
Timing = Placement and routing answer to timing requirements
SIZE = Minimum size possible
None = Normal random logic
NOTE: This timing does not include verification time.

- Size limitation
- Routing layers available—for example, only three out of five may be a limitation

Table C.2 is a block-level schedule example.
Let's try now to define the factors that affect scheduling for a full chip:

- *Experience level of team*: Does the team have enough experienced people?
- *Change*: Evaluate the risk of certain key parameters changing over the project. For example, pad positions and even design rules are subject to change over the course of a long project.
- *Reuse*: Can we leverage experience and layout designs that were done before?
- *Design complexity*: Number of critical blocks, signals and/or busses.
- *External factors*: Is the team colocated or is it a joint design project with outsiders?
- *Third-party blocks*: How easy will it be to import a block from an intellectual property (IP) provider?
- *Methodologies*: Are there any new flows and/or tools that have to be introduced?
- *CAD support*: Does the project team get CAD support? What is the priority of the project?
- *Team size*: A large team may not be as productive as a small one because of communication and management overhead.
- *Work day*: Is overtime assumed or planned for?
- Sick leave, bereavement, vacation time, seasonal restrictions.

Project scheduling is an art and a science, and the preceding list is intended to give you a feel for the complexity of the task.

INDEX

About the CD-ROM...

Color Art Examples

On the CD, you will find color versions of many of the figures from the book. These more complex color examples have been included here to help you better understand the concepts demonstrated by the printed black and white versions.

Some of the complex chips included on the CD are the property of MOSAID Technologies, Inc., who agreed to let us include them in order to help the reader better understand and evaluate different types of layout design combined in a real environment. Check out the **Flows** and **Cell Library** presentations. You can view them within your web browser, in their native PowerPoint format, or as Acrobat pdf files.

Presentations

During the development of the text we found that there are some very "hot" issues today, such as extraction of layout resistance and capacitance, migration tools and principles, deep submicron and very deep submicron design, etc. We decided to talk with marketing managers from Mentor Graphics, Cadence, Sagantec, and others and asked them to include some of their presentations addressing these new concepts, methodologies, and tools.

The **Cadence Design Systems** presentation talks about ASIC flow and demonstrates how physical information can be used earlier in the design process in order to provide designers with "real world" data for simulations.

The **Mentor Graphics** presentation emphasizes the extraction importance in 0.25 microns and below gate-size processes.

The **Sagantec** presentation talks about the concept of migrating layout from one process to another, compaction within a layout polygon editor environment, and a new concept called "enlargement" or reverse compaction for wires.

There are many other tools and presentations available on various Electrical Design Automation (EDA) vendors' sites, but these are the only ones that answered our request in time for the publishing deadline. We are confident that for a future revision of our book, more EDA vendors will provide interesting topic presentations without too much "tool sales pitch."

Software

The most incredible contribution to this book comes courtesy of Tanner Research, Inc., who put Dan's library inside their **Tanner Tools L-Edit demo**, which is included on our CD-ROM. This tool offers you a tremendous opportunity to practice as you learn about each topic.

Most the files on the CD have been prepared as web-ready (htm or gif), Adobe Acrobat (pdf), and PowerPoint slideshows (pps). We've included these files in multiple formats to offer you greater viewing access than a single format would allow.

Accessing the CD Contents

To surf the contents of the CD, you will need to have a web browser installed on your computer. If you already have a web browser installed, launch the application and open the file "**D:\Readme.htm**" (where "**D**" is the designation of your CD drive).

If need be, you can install Microsoft Internet Explorer 5.0 directly from the CD by running "**D:\Software\IE5\IE5Setup.exe**". Follow the directions on screen to complete the installation. After successfully installing the web browser, launch the application and open the file "**D:\Readme.htm**" (where "**D**" is the designation of your CD drive).

All of the contents have been hyperlinked within **Readme.htm**. To access any file directly, simply point and click to the file you'd like to view or the software you wish to install.

Technical Support

Beyond providing replacements for defective discs, Butterworth-Heinemann does not provide technical support for the software included on this CD-ROM.

Send any requests for replacement of a defective disc to Newnes Press, Customer Service Dept., 225 Wildwood Avenue, Woburn, MA 01801-2041 or email **techsupport@bhusa.com**. Be sure to reference item number **CD-71947-PC.**

File Directory ...

Directory\File Location	File Description
Readme.txt	Basic CD information
Readme.htm	Main CD interface. Open this file within your web browser to link to the main CD contents
\Art\Art.pdf	All color figures contained within one Adobe Acrobat file.
\Art\Figure_2-7.gif	Color version of printed book figure.
\Art\Figure_2-10.gif	Color version of printed book figure.
\Art\Figure_2-12.gif	Color version of printed book figure.
\Art\Figure_3-2.gif	Color version of printed book figure.
\Art\Figure_3-3.gif	Color version of printed book figure.
\Art\Figure_3-4.gif	Color version of printed book figure.
\Art\Figure_3-5.gif	Color version of printed book figure.
\Art\Figure_3-6.gif	Color version of printed book figure.
\Art\Figure_3-8.gif	Color version of printed book figure.
\Art\Figure_3-9.gif	Color version of printed book figure.
\Art\Figure_3-10.gif	Color version of printed book figure.
\Art\Figure_3-11.gif	Color version of printed book figure.
\Art\Figure_3-12.gif	Color version of printed book figure.
\Art\Figure_3-15.gif	Color version of printed book figure.
\Art\Figure_3-16.gif	Color version of printed book figure.
\Art\Figure_3-17.gif	Color version of printed book figure.
\Art\Figure_3-18.gif	Color version of printed book figure.
\Art\Figure_3-19.gif	Color version of printed book figure.
\Art\Figure_5-2.gif	Color version of printed book figure.
\Art\Figure_5-2A.gif	Color version of printed book figure.
\Art\Figure_5-6.gif	Color version of printed book figure.
\Art\Figure_5-12.gif	Color version of printed book figure.
\Art\Figure_5-13.gif	Color version of printed book figure.
\Art\Figure_5-14.gif	Color version of printed book figure.
\Art\Figure_5-15.gif	Color version of printed book figure.
\Art\Figure_5-26.gif	Color version of printed book figure.
\Art\Figure_5-28.gif	Color version of printed book figure.
\Art\Figure_5-29.gif	Color version of printed book figure.
\Art\Figure_7-3.gif	Color version of printed book figure.

Directory\File Location	File Description
\Art\Figure_7-8.gif	Color version of printed book figure.
\Art\Figure_7-20.gif	Color version of printed book figure.
\Art\Figure_7-22.gif	Color version of printed book figure.
\Art\Figure_7-26.gif	Color version of printed book figure.
\Art\Figure_7-27.gif	Color version of printed book figure.
\Art\Figure_8-6.gif	Color version of printed book figure.
\Art\Figure_8-7.gif	Color version of printed book figure.
\Art\Figure_8-13.gif	Color version of printed book figure.
\Art\Figure_8-16.gif	Color version of printed book figure.
\Presentations\Cadence\Cadence.htm	Cadence presentation saved as a web-ready htm.
\Presentations\Cadence\Cadence.pps	Cadence presentation saved as an independent PowerPoint file.
\Presentations\Cadence\Cadence.exe	Cadence presentation packed with a PowerPoint viewer.
\Presentations\Cell Libraries\Cell Libraries.htm	MOSAID Cell library presentation saved as a web-ready htm.
\Presentations\Cell Libraries\Cell Libraries.pps	MOSAID Cell library presentation saved as an independent PowerPoint file.
\Presentations\Cell Libraries\Cell Libraries.exe	MOSAID Cell library presentation packed with a PowerPoint viewer.
\Presentations\Cell Libraries\Cell Libraries.pdf	MOSAID Cell library presentations saved as an Adobe Acrobat pdf.
\Presentations\Flows\Flows.htm	MOSAID Flows presentation saved as a web-ready htm.
\Presentations\Flows\Flows.pps	MOSAID Flows presentation saved as an independent PowerPoint file.
\Presentations\Flows\Flows.exe	MOSAID Flows presentation packed with a PowerPoint viewer.
\Presentations\Flows\Flows.pdf	MOSAID Flows presentations saved as an Adobe Acrobat pdf.
\Presentations\Mentor Graphics\Mentor Graphics.htm	Mentor Graphics presentation saved as a web-ready htm.
\Presentations\Mentor Graphics\Mentor Graphics.pps	Mentor Graphics presentation saved as an independent PowerPoint file.
\Presentations\Mentor Graphics\Mentor Graphics.exe	Mentor Graphics presentation packed with a PowerPoint viewer.
\Presentations\Sagantec\Sagantec.htm	Sagantec presentation saved as a web-ready htm.
\Presentations\Sagantec\Sagantec.pps	Sagantec presentation saved as an independent PowerPoint file.
\Presentations\Sagantec\Sagantec.exe	Sagantec presentation packed with a PowerPoint viewer.
\Software\Acrobat Reader 4.0 + search\CD\Reader\ArcrRd32.exe	Adobe Acrobat Reader.
\Software\Acrobat Reader 4.0 + search\CD\Install\rs40eng.exe	Adobe Acrobat Reader installation program.
\Software\IE5\ie5setup.exe	Microsoft Internet Explorer 5.0.
\Software\Tanner Tools\Setup.exe	Tanner Technologies Tool demo.